Second Edition

RESEARCH
METHODOLOGY
IN APPLIED
ECONOMICS

Second Edition

RESEARCH METHODOLOGY IN APPLIED ECONOMICS

Organizing, Planning, and Conducting
Economic Research

Don Ethridge

Blackwell
Publishing

Don E. Ethridge, Ph.D., is Chairman and Professor, Department of Agricultural and Applied Economics, Texas Tech University, Lubbock.

©2004 Blackwell Publishing

Blackwell Publishing Professional
2121 State Avenue, Ames, Iowa 50014, USA

Orders:	1-800-862-6657
Office:	1-515-292-0140
Fax:	1-515-292-3348
Web site:	www.blackwellprofessional.com

Blackwell Publishing Ltd
9600 Garsington Road, Oxford OX4 2DQ, UK
Tel.: +44 (0)1865 776868

Blackwell Publishing Asia
550 Swanston Street, Carlton, Victoria 3053, Australia
Tel.: +61 (0)3 8359 1011

Printed on acid-free paper in the United States of America

First edition, © 1995, Iowa State University Press
Second edition, © 2004

Library of Congress Cataloging-in-Publication Data

Ethridge, Don E. (Don Erwin)
  Research methodology in applied economics: organizing, planning, and conducting economic research/Don Ethridge—2nd ed.
    p. cm.
  Includes bibliographical references and index.
  ISBN 0-8138-2994-1 (alk. paper)
  1. Economics—Research. 2. Economics—Methodology. I. Title.

  HB74.5.E84 2004
  330'.9072—dc22   2004001461

The last digit is the print number: 9 8 7 6 5 4 3 2 1

Contents

Preface   ix

I. Research Methodology Foundations

1. Introduction   3
Reasons to Study Methodology   4
Objectives and Focus of the Book   7
Assumptions about Readers   9
The Author's Perspective and Views   10
Organization of the Book   13
Summary   14
Recommended Reading   14
Notes   14

2. Research and Methodology   15
Research Defined and Described   15
Classifications of Research   19
    Basic vs. Applied Research   20
    Disciplinary, Subject-Matter, and Problem-Solving
        Research   20
    Analytical vs. Descriptive Research   24
Methodology Defined and Described   25
The Process of Research   28
Creativity in the Research Process   30
Summary   33
Recommended Reading   34
Suggested Exercises   34
Notes   35

**3. Methodological Concepts and
     Perspectives   36**
Science   36
Economics as Art and Science   38

Knowledge   40
    Positivistic vs. Normativistic Knowledge   40
    Private vs. Public Knowledge   42
Ways We Obtain Knowledge   43
Reliability of (Public) Knowledge   45
Logical Fallacies   46
Tests for Reliability   47
Role of Personal Objectivity   49
Scientific Prediction   50
Summary   55
Recommended Reading   56
Suggested Exercises   56
Notes   56

4. **Philosophical Foundations   58**
Positivism   60
Normativism   63
Pragmatism   66
How the Philosophies Blend   68
Empiricism in Research Methodology   71
The Scientific Approach   72
    Deduction and Induction in the Scientific
    Approach   74
Summary   77
Recommended Reading   78
Suggested Exercises   78
Notes   78

II. The Research Project Design

5. **Planning the Research   83**
The Research Project Proposal   84
Elements of the Research Proposal   85
Evaluating Research Proposals   89
Importance of Flexibility   90
Funding for Economic Research   91
Importance of Writing   92
Writing Guidelines and Tips   94
Summary   98
Recommended Reading   98
Suggested Exercises   98
Notes   99

6.   The Research Problem and Objectives   100

Research vs. Decision Problems   101
The Research Problem Identification   102
Researchable Problem vs. Problematic Situation   104
The Problem Statement(s)   105
Use of Data   107
Objectives   108
Developing Problem and Objective Statements   109
Summary   111
Suggested Exercises   112
Notes   112

7.   The Literature Review   113

Purposes of the Literature Review   113
The Literature Search Process   115
   Search Aids   116
   Key Words   117
   Reading   118
   Notes   118
Writing the Literature Review   119
Referencing   120
Summary   124
Recommended Reading   125
Suggested Exercises   125
Notes   126

8.   The Conceptual Framework   127

Role of the Conceptual Framework   129
Source Material for the Conceptual
   Framework: Theory   131
Hypotheses and Hypothesis Testing   132
Operational Suggestions   136
Summary   138
Suggested Exercises   139
Notes   139

9.   Methods and Procedures   140

Purposes of Methods and Procedures   141
A Historical Perspective on Empirical Methods   142
Models in Economic Research   144
Types of Empirical Methods   146
   The Descriptive Method   147
   Statistical and Econometric Tools   147
   Operations Research Tools   148

Data Considerations   150
    Secondary Data   150
    Primary Data   151
    Creative Solutions for Data Problems   152
Procedural Suggestions   153
Summary   154
Recommended Reading   154
Suggested Exercises   155
Notes   155

III. Closure of the Research Process

10. Reporting the Research   159
Types of Reports   160
Components of the Research Report   162
Writing the Methods/Procedures and Findings   165
Writing the Conclusions   167
Publishing   168
Authorship   170
Summary   172
Recommended Reading   173
Suggested Exercises   173
Notes   173

IV. Appendices

Appendix A. Example of a Research Proposal for a
    Government Agency   177

Appendix B. Example of a Master's Thesis
    Proposal   190

Appendix C. Example of a Ph.D. Dissertation
    Proposal   201

Appendix D. Guidelines for Critiquing
    Papers   229

Appendix E. Seeking Research Funding   231

References   236

Index   242

Preface

The premises of this book are that there is need for formal, focused attention within the economics discipline on how to organize and conduct research in economics and that we can increase the effectiveness and efficiency of our research process by formal study of it. The writing of the book was undertaken because of a perceived need for a graduate-level textbook that provided more insight into the procedural aspects of research methodology in economics—how to approach and organize the process of conducting applied research in economics or to fill gaps in the instructional material available to students and teachers of the subject. The second edition remains consistent with that theme but adds clarification and explanation to the conceptual and philosophical dimensions of research methodology. The book is intended to represent a core or synthesis of the subject matter of research methodology in economics as the author understands it, not the final word on the subject.

This edition was undertaken because of the growing conviction that it was needed. I was encouraged by colleagues using the book in their efforts to teach some of the same principles and by the belief that several aspects could be improved. Over the course of my career as an economist there has been a gradual, persistent decline in understanding of the research *process* by the rank and file of economists. We have, as a group, become more sophisticated with a portion of the process, particularly the technical methods and procedures, but a large contingent of economists and economics students lack comprehension/understanding of the *entire* process. I am compelled to try to improve that situation, and this book represents that attempt. I hope it helps make a difference.

My approach is to go to the fundamentals, some of which are decades old, some of which are centuries old, and some of which are current. I am not alone in the attempt to improve understanding of the process, which should be apparent to those who examine the citations and references used. The fundamentals of research methodology draw a clear distinction between *methodology* and *methods*, which much of the current professional literature and journal authors fail to discern. Good methodology requires good methods (techniques), but it also requires good and thorough *thinking*, comprehension of the *science* and the research *subject matter*, good *writing*, and persuasive *interpretation* (and argument). I believe that for us to be relevant economists, we must apply the process diligently because it is the most we can do to ensure that the knowledge/information we produce is the best it can be.

As you read the material in this book, I ask that you first, consider the interpretations, descriptions, thoughts, and prescriptions in the context of their overall meanings, not in bits and pieces. For example, consider the definition of *research* in the context of the entire discussion of the research process rather than merely a one-sentence definition that might be found in a dictionary. The subject matter is too complex to reduce to simplistic statements, although synthesis can be quite valuable in the process. Second, I ask that you read other sources; overall, you will better understand the positions, prescriptions, and descriptions if you augment this material with related materials, and on the points on which we disagree you will better understand the basis for our differences.

The two central objectives of the book are to provide (1) an overview of the conceptual and philosophical basis of research methodology in economics; and (2) procedural guidelines on planning, designing, and conducting research projects. The emphasis is on *research* methodology as it applies to economics, not economic methodology, philosophy of science, or research tools and techniques, although research methodology overlaps each of these areas.

The book is divided into three parts: the definitional, conceptual, and philosophical aspects of research methodology (chapters 1–4); the procedural aspects of research methodology (chapters 5–9); and reporting of the research (chapter 10). Chapters 2–4 define terms and concepts, examine and discuss the process, and consider how research and methodology are related to science, knowledge, objectivity, prediction, induction, deduction, and different philosophical beliefs. These chapters provide understanding, rationalization, and justification of the process as it is presented in subsequent chapters.

Chapters 5–9 are devoted primarily to planning and organizing a research project. The section starts with an overview of the research plan, which in written form is the research proposal, then proceeds to major elements of the proposal. Substantive procedural issues on matters such as suggestions and perspectives on writing, seeking funding support for research, and data collecting are included, either within the chapters or as appendices. Chapter 10 deals with reporting research rather than planning and organizing it.

Some chapters include a list of suggested reading that add to or support material in the book. These recommended sources, while not the only relevant sources, augment the book or may be particularly useful or thought-provoking for readers in their own exploration of the subject. Some recommended exercises have also been included at the end of chapters as devices to help students in their efforts to integrate the concepts and guidelines into their own thought processes and approaches. There is no better way to grasp the principles than to gain experience using them.

An attempt has been made to present the material such that the book does not necessarily have to be covered in the same order of the progression of chapters, although that progression may be best for a long-term study of the subject. For example, one might cover chapters 1 and 2, then 5 through 9, then 3 and 4, then 10; this sequence is especially helpful when students are developing a research proposal as a requirement of a course, because earlier focus on procedural aspects helps in initiating the proposal earlier in the term of the course. Also, the order of coverage could be altered within major divisions of the book. For example, within section II (procedural), the order could be changed to 5, 7, 8, 6, and 9; this approach could serve to emphasize that the problem definition, objectives, literature review, and conceptual framework must, in the final analysis, exist as an integrated body of thinking rather than separated components. The research process is too complex and requires too much creativity to be reduced to a simplistic recipe.

An attempt has also been made to present the material in a way that portions may be foregone when situations dictate. For example, chapters 3 and 4 might be omitted for undergraduate students when the intent is to place the entire emphasis on research tools/techniques (not an approach endorsed by the author). Emphasis on the conceptual framework can be overdone with undergraduates because they are unaccustomed to reasoning with theory. The author's prescription in these cases is to utilize the conceptual framework but maintain the theoretical sophistication at the level of students' training. These considerations aside, the recommended use of the book is as an integrated whole.

I am indebted to many people for various contributions in bringing this work to its present state. At the risk of overlooking someone, I must mention several of them; each has helped me in developing this book, but none bears any responsibility for its shortcomings. Three people have impacted my foundational thinking about the subject: the late Willard F. Williams introduced me to the field of study and served as both mentor and colleague; Glenn L. Johnson patiently gave much time and energy to help me understand the philosophical positions ingrained in science and how they influence the research process; and George W. Ladd provided new insights for me, particularly on how research philosophies and the practice of research influence one another. Many who contributed to the first edition of the book are acknowledged in the preface of that edition and are not repeated here. In this edition, several have made especially notable contributions. First among those are Darren Hudson, whose perspective from using the first edition and thoughtful comments, opinions, and questions on various subjects have been most helpful. I extend special thanks to Anne Macy Terry (Anne Marie Macy in appendix C) and Shombi Sharp for permitting me to use their graduate program proposals in the appendices. Helpful comments and suggestions were also provided by Larry Janssen, Thomas Rutherford, and several anonymous reviewers. The editorial staff at Blackwell Publishing was most helpful and efficient, and I want to express a special thank-you to Gretchen Van Houten for her assistance and support in bringing the book to publication. Finally, I am grateful to the scholars and authors whose work has provided perspective and insight for me and to the substantial number of students and colleagues who have made me think about the matters that are the subjects of this book.

On a very personal note, this book is dedicated to my brother, Dean Ethridge, also an economist, from whose courage and character came a gift and, consequently, the motivation to undertake this edition. He will disagree with at least some of my interpretations and prescriptions, but that will provide us with opportunities for argument and debate—activities that we have both enjoyed for most of our lives.

I
Research Methodology Foundations

1

Introduction

We do a better job of teaching theory and tools than . . . teaching their use. . . . Students need more training in the relevance and application of the tools.
Commission on Graduate Education in Economics,
American Economic Association, 1991

Economic research, in its most fundamental form, is the process by which we discover, evaluate, and confirm the stock of economic knowledge. That knowledge base consists of disciplinary economic knowledge, subject-matter knowledge of the economic components of many different subject-matter areas, and problem-solving knowledge on matters to which economics has something to offer. While the focus of this book is on *applied* economics—using economic concepts and tools to address a full range of contemporary issues—*disciplinary* economics is clearly relevant to that purpose, since it is the source of the fundamental concepts and tools that are necessary for applying economics. The meanings of these statements should become clearer as the contents of the book unfold.

Our capacity to accomplish economic research, at least certain types of it, is expanding so rapidly that few individual economists can keep pace with it. This is particularly the case with, and is most obvious in, our capability to process and evaluate data and use quantitative tools. Computer technology lies at the heart of these advancements. This technology presents us simultaneously with great opportunities and great risks, depending on how we manage and exploit the technology and its capabilities.

Computer technology, in combination with increasingly sophisticated statistical and mathematical simulation techniques, gives us the means to explore and explain the complex nature of economic phenomena and relationships, many of which have been previously unobtainable. Yet the ease with which we can process information and solve complex mathematical formulations also lends itself to abuse and may tempt us with false confidence about the knowledge we possess. A computer model with erroneous internal reasoning can produce misinformation

3

as rapidly as one programmed correctly can produce valid information. You may recall the adage, "Figures don't lie, but liars figure." Consider the question asked of a professor in an econometrics course as the class was discussing less than completely specified models. Someone asked, "Isn't some information better than no information?" He replied, "Is wrong information better than no information?" Just because we can grind through masses of numbers does not ensure that the results from the process will be correct, useful, valid, or trustworthy.

Doing research is an activity through which we expand our understanding of the world about us. We use that understanding or knowledge for varied purposes. If the knowledge is to be reliable, the research must be done with care and planning. Reliability of information rests on its validity and applicability, never on what the information turns out to be. We affect validity and applicability of information by the processes and procedures through which it is generated. It is not data (facts) that determine whether our knowledge is reliable. Facts may lend themselves to many different interpretations. Analytical tools or techniques alone cannot make our knowledge reliable, because the analytical tools have no capacity to do anything other than what we specify. Theories alone cannot make our knowledge reliable. Our theories are deductive abstractions about conditions that relate only to selected situations. None of these in isolation has the capability to provide us with knowledge. Yet all of them together, applied in a reasoned, deliberate manner, but with an open-minded, questioning attitude, have the capacity to provide reliable knowledge and are the basis of valid science.

This book is prompted by a perceived need for better understanding of this process—and the craft of accomplishing it.

REASONS TO STUDY METHODOLOGY

Methodology refers to the *manner* in which we approach and execute functions or activities. It constitutes general approaches or guidelines, not specific details of how we accomplish a task. The specifics are the *methods* (and/or procedures), not methodology. Within a discipline, methodology constitutes the agreed-upon or accepted rules of evidence for reasoning about the various aspects of that discipline. Consequently, economic methodology and research methodology in economics are not precisely the same, although there is a substantial impact of economic methodology on the more specific subject-matter area of research methodology in economics. Research methodology provides the principles for organizing, planning, designing, and conducting research, but it cannot tell you in detail how to conduct a specific, individual research study. Each study is unique.

> The central reason for studying research methodology is that it provides
> a time-tested, proven means of producing new, reliable knowledge. That
> accumulated, growing body of knowledge comprises a discipline, or per-
> haps a "science."

Research methodology in economics, and the study of it, is the integration of the various components of the study of economics in an effort to accomplish a defined, goal-directed function—research. Economic research methodology is, in a sense, the science and art of economics rolled together, with its goal being both to expand (and confirm) our knowledge and to make that knowledge relevant to the contemporary world. Research methodology is not peripheral to the primary thrust of the study of economics; it lies at the very heart of economics. However, economics also encompasses activities other than research. Besides education, these activities extend to advising governments, private industries, and individuals in decision making and the implementation of policies.

Our approach to education and training within academic disciplines is to segment the discipline into subject-matter groupings (courses) so that we can manage the material within time and other resource constraints. In economics we have general theory courses (e.g., macroeconomics, microeconomics, welfare theory) and subject-matter courses (e.g., resource economics, labor economics, international trade). We also delve into related disciplines (e.g., mathematics, statistics, management, political science), analytical courses (e.g., econometrics, simulation, operations research methods), and so on. This approach to learning is efficient in that it allows us to segment and construct programs from combinations of components (individual courses). However, we approach learning to do research much the same as we approach learning to teach—by assuming that we somehow understand how to do it by example. It is not surprising that, as Ladd (1987, p. 5) points out, students are typically surprised and troubled to find that their research does not proceed in the easy, organized way that their reading of journal articles leads them to anticipate.

This does not suggest, however, that learning to do research by example is not the way to learn. In fact, the contention here is that the only way we learn how to do research is to *do* it. The process of doing research under the supervision of an advisor in the tradition of a master-intern relationship is an effective model. However, formal instruction on how to organize and conduct research can make the process more effective for both master and intern. Formal instruction in research methodology is not common in economics programs. Isolated elements or parts of research methodology are often covered in other courses but are not frequently pulled together for concentrated consideration, examination, study, and application except in a subset of applied economics programs. Instead of research methodology, courses in many economics programs typically deal with economic methodology, development of economic thought, and/or philosophy of science, although focus on these areas has declined in recent years, yielding to quantitative "tools" courses. The subject matter of research methodology is an integral part of being able to pull together our fragmented or segmented attention on the various methods and techniques of our discipline into a more coherent overview of them. A broader perspective on research methodology is, in turn, necessary to advance the state of understanding within the discipline. We can increase the effectiveness and efficiency of our research process by formal study of the process.

Many in economics and in other disciplines appear to agree. McCloskey (1990a, p. 1127), for example, indicates that the methodological thinking of economists as it relates to science and research is in general "a scandal." A report by the Commission on Graduate Education in Economics for the American Economic Association reported that graduate training does little to develop the skills of graduate students for the actual conducting of research (Hansen, 1991). They receive in-depth technical and theoretical training, but they are too frequently incapable of applying the concepts and tools to contemporary situations. Two of the commission's major recommendations for improvement of graduate programs in economics in the United States are (1) more emphasis on "real-world" problems and the application of economic research to them; and (2) more emphasis on communication skills, especially writing, and the ability to relate economic knowledge to the public (Krueger et al., 1991). Furthermore, conducting research under proven guidelines and with supervision is the best way to learn to do useful economic research (Hansen, 1991, p. 1079). Colander (1998) addressed the commission report from the perspective of the direction of research and focus of resources within graduate economics programs. He noted that working with formal economic theory requires a "formalist deductivist" mind-set, while doing applied creative research requires a "generalist inductivist" perspective (p. 605). He maintains:

> Technical applied work is the most difficult branch of economics. To do it well, one must be a good generalist and a good formalist. . . . Unless graduate economics programs come to grips with the need for applied work, they will not only fail to grow, as has been the case in the past 20 years, but will shrink in size (pp. 605–606).

A colleague expressed the shortcoming of our educational programs in another way: We teach students to *know* economics, but we rarely teach them to *do* economics.[1]

These are critical areas in which current economics programs are not measuring up to the standard judged to be desirable or necessary within the profession. The Committee on the Conduct of Science of the National Academy of Sciences likewise supports the need for formal training in research methodology, indicating that beginning researchers working with experienced scientists is one of the most important aspects of scientific education, but that this informal structure is not enough (Ayala et al., 1989, p. v). Economics graduate students also appear to understand that various aspects of their education are out of balance. Colander and Klamer (1987) report from a survey of U.S. students in economics graduate programs that students perceive that their success in graduate school is determined primarily by their understanding of techniques: success has little to do with understanding the economy or economic literature, or being able to conduct economic research.

There is also a lack of understanding of the research process among some professional economists. The need for formal understanding does not magically disappear once a graduate degree is granted. There are practicing professional economists who do not know how to *do* research. And just getting journal articles published is not, in itself, evidence that one knows how to do research: in the author's experience, based on reviewing a substantial number of papers for editors of various journals, well over half of them demonstrate some fundamental methodological flaw in their initial submissions. These manuscript flaws are frequently not errors in empirical or mathematical analytical technique. Common problems are failure to establish a reason for the research (the problem) or to provide a clear, concise understanding of research objectives. Failure to provide complete referrals to relevant prior research literature and/or lack of attention to the conceptual or theoretical basis of the research are not rare oversights. Selection of analytical structural models for mere empirical convenience and presenting conclusions that are no more than restatements of analytical findings are also prevalent problems. It is not possible to tell what proportion of these "flaws" is due to problems of understanding of the process and what is due to communication problems. However, both types of problems indicate, as suggested by the Commission on Graduate Education in Economics (Krueger et al., 1991), a substantial problem in the profession.

OBJECTIVES AND FOCUS OF THE BOOK

This book is intended to increase the reader's proficiency and effectiveness in economic research. It is designed for use primarily by graduate students in all types of economics programs. It may also have application in some analytically oriented undergraduate economics programs for advanced students and as a basic reference for some professional economists.

The two primary objectives of the book are to provide (1) an introduction to the conceptual and philosophical foundations of research methodology as it applies to research in economics; and (2) procedural instruction on how to plan, design, and conduct research projects in economics.

The goal of the book is to provide understanding and instruction on how to do economic research. Doing research entails planning and designing the research, implementing and completing the analysis, and disseminating the results of the research. Students are especially vulnerable to numerous pitfalls of "bad" methodology. Examples include initiating the process without a clear understanding of the research problem and/or objectives or initiating research before knowing the related research that has already been done. Among many other errors are inadequate or incomplete conceptualization of the problem and confusing research means and ends. Good research—research that is defensible, useful, and expands our knowledge base—is no accident.

This book is not about the philosophy of science, although selected components of alternative philosophies that are particularly germane to the philosophical foundations of research methodology are summarized. Nor is it about economic methodology—the approach to economic reasoning—although components of economic methodology are obviously important. Aspects of economic methodology are covered but are not the primary concern. The book is likewise not about research methods—the tools and techniques employed in economic analysis—although how we select and use individual techniques or tools, or combinations of them, is also of concern in methodology. While each of these specialized subject-matter areas is relevant to students of economics, our purpose is to concentrate on *research* methodology in economics, or the study of how to approach and do all types of economic research. Furthermore, heavier emphasis is placed on "applied" research in economics, where there appears to be a clear need for students to focus attention and gain experience and perspective. It is also maintained, as discussed in chapter 5, that the greater demand by users of economic research is for "applied" research and that our "basic" research tends to be in part influenced by those activities.

The diverse nature of the field of economic study is also worth noting. Economics as a discipline touches almost all human activities, yet the picture it provides of these activities is incomplete. Our economic activities are affected by social, religious, political, governmental, and all manner of institutional factors, as well as physical, biological, and other constraints. We are inherently multidisciplinarians to some degree; we must be to be proficient, relevant, applied economists. Consequently, much of our research is, of necessity, multidisciplinary and oriented toward problem solving. Within universities, we work in colleges of arts and sciences, business administration, agriculture, engineering, and others. We fit everywhere, yet we fit nowhere easily. Still, some maintain that we may have carried "super-specialization" too far, thus failing to capture the gains from trading intellectual concepts with other disciplines (McCloskey, 1990a, pp. 1127–1129).

This book addresses two divergent, but related, aspects of economic research methodology that do not appear to have been previously addressed together.[2] These are processes of discovery and confirmation. The process of confirmation deals with discerning the validity or reliability of knowledge or information. The process of discovery deals with formulating, finding, and creating new knowledge or information. They are inherently different, but both are intertwined and obviously important to research and science. Discovery is inherently a creative process (an "art" side of any discipline), requiring questioning, probing, pursuing alternative avenues of exploration, and the like. While essentially creative, it benefits substantially from a degree of structure, and a considerable portion of this book focuses on that aspect. Confirmation, determining the reliability of knowledge, is more highly developed and the procedures are more widely agreed upon;

this is the "science" side of a discipline. A discipline cannot confirm without discovery, but it can discover without confirming.

Many of the books on philosophy of science are effectively philosophies of confirmation. Fewer books exist on the philosophical aspects of the methodology of economics, and Glenn Johnson's book, *Research Methodology for Economists* (1986b), is among the best. Useful writings on the process of discovery in economic research are less focused in a formal sense, and virtually none are comprehensive, but many of these sources have been used extensively in this book. The point of this explanation is to identify the goal of including both processes—discovery and confirmation, art and science—in a broader perspective of research activity as it is applied within disciplinary and applied economics. The degree to which the goal is met will be judged with the perspective of hindsight.

ASSUMPTIONS ABOUT READERS

In addressing the topic, some assumptions are necessarily made about the background, experience, and mental attitude of readers. It is assumed that you are a serious student of economics and are motivated to engage in research on matters with economic content. It is also assumed that you already possess a working knowledge of general macroeconomic and microeconomic theory and several subject-matter fields within economics. You are assumed to have basic skills in the techniques of statistics, econometrics, and operations research procedures, plus sufficient exposure to other disciplines and disciplinary perspectives to relate well to the larger world of which economics is a part. This general level of academic exposure is desirable, perhaps necessary, for you to grasp the generalities of methodology as it applies to applied economic research, which is inherently multidisciplinary to some degree. It is assumed that you can think abstractly, but that is probably redundant. If you have survived the above types of academic exposure, it should constitute evidence of your ability to reason in the abstract.

There are other attributes that you need in order to address research methodology as well as other subject-matter areas. One is critical thinking. If you question and examine what you read and are told, you will gain more from it. You may not accept or agree with all aspects of research methodology that are presented here. However, you really cannot agree or disagree with what is presented if you do not consider it thoroughly and critically. By critically considering it you will internalize that part which is useful to you (i.e., make it a part of your own thinking processes). On the other hand, critical thinking carried to the extreme may become cynicism, which often becomes a barrier to understanding.

The ability to synthesize is a very important characteristic of economic research. This includes the capacity to pick and choose from among many concepts, facts, and prior research and select and combine the elements that are relevant to a particular problem, question, or issue. Understanding various theories and facts is a necessary, but not a sufficient, condition for synthesizing. The synthesis extends to recognizing what parts of those theories and facts relate to the research task at hand. The ability to synthesize improves with practice and experience, but we cannot impart the basic capacity by teaching: some have it or can develop it, others cannot. However, if the ability exists, we know of some things that can increase the capacity. For example, Ladd's (1987) suggestions for being more creative as described in his chapter 2, "Creativity in the Process," can foster the ability to synthesize. The basic capacity for synthesizing must exist for you to plan and organize research, and assuming that you possess that characteristic, you will need to invest in improving it. If you cannot synthesize, the usefulness of this book to you will be severely limited.

Another attribute that will serve you well as you study the subject is the mental and emotional maturity to discern the difference between a privately held belief or value and a result or concept that is supported by evidence. This ability fosters, although it does not guarantee, the maturity to come to grips with the difference between how things are and how we may wish they were. This maturity does not automatically occur with age, although age *may* improve it if it already exists. We do not know what it is made of, but it is associated with mental discipline and an objective state of mind.

My experience has convinced me that students should study research methodology while engaged in a research process. You can read and hear about how to plan and conduct research, but you will not begin to comprehend the meanings within the material until you have put them into practice. Perhaps this is true with all other subjects as well. The best way for you to understand the principles and procedures provided in this book is to embark on your research work (the planning and implementation) in conjunction with your research advisor(s), either while studying the material in this book or immediately thereafter. I also believe that you will not grasp the full meaning of the material of this book by reading it only once, or by reading this book apart from reading other material on the subject. Many of the precepts and guidelines presented in this book will become clearer after you begin to accumulate research experience.

THE AUTHOR'S PERSPECTIVE AND VIEWS

Much of the material of this book—thoughts, ideas, concepts, perspectives, prescriptions, guidelines—is not unique or original to the author. I have read, studied, learned, integrated, adapted, and synthesized most of it from the works of others. Yet what I present to you and the way I present it are unavoidably colored by my individual interpretations, perspectives, and interests. I therefore consider it necessary to attempt, as best I am able, to forewarn you about my biases—the way I am predisposed to perceive and examine things. I am not very confident that I understand my own biases with a high degree of clarity or objectivity. However, I am sure that elements of my background, interests, and beliefs provide some insight into those biases. I attempt to provide you with some relevant information with which to help you consider the contents of this book.

Having been a student of economics for about forty years, I am still fascinated by economics. When I ponder a question or issue with economic content—and most questions and issues have economic content—my first reaction is to place it into a theoretical or conceptual context: that is, to mentally search for the economic concept or combination of concepts that provides insight into the question or issue. On the other hand, I tend to focus on applied subject-matter and problem-solving types of questions and issues. To me, the beauty of economics rests in its theory, but the power of economics lies in its application to current problems. There is a strong influence of pragmatism in my philosophical approach to economics, along with a conviction that it is our theory that lies at the center of whatever ability we have to make sense of the world we study and observe and to solve problems. I see applied and basic research as separable in a definitional sense, but very intertwined in practice.

My philosophical beliefs are, in my perception, a mixture of positivism, normativism, and pragmatism, but relatively heavy on pragmatism. Some colleagues view me as very pragmatic, yet others see me as quite positivistic. On the other hand, I have frequently engaged in policy research/analysis, which I believe is very important, but which is unavoidably normativistic. But that is one of the interesting things about private beliefs—we tend to classify others' positions in relation to our own. I see economics as both an art and a science, depending on its specific application in a particular circumstance. I approach economics as an analytical discipline, with no relation to a single political philosophy. In fact, mixing political positions or philosophy with economic analysis often makes a mess of both of them in my view.

I have little patience with a "school of thought" mind-set. Adherence to any line of reasoning to the exclusion of another seems dogmatic and unproductive. I also do not accept notions or assumptions of a "hierarchy" of research types, topics, or approaches. That is, "basic" research is not superior to "applied" research (or vice versa), disciplinary research is not superior to multidisciplinary research, research done in academic institutions is not superior to research done in government or private industry, and so on. Each research effort must be judged in the context of its intended purpose, its contribution, and the quality of its content. Some of the "best" research may achieve its intended purpose by being simple and direct. Much of our best research is also not publishable in our disciplinary and subject-matter journals because it is either too applied or too intricate in its construction to fit within journal format.

I am an advocate of quantitative methods in economic research. I believe that they lie at the heart of whatever capacity we as a profession have to obtain meaningful understanding of economic phenomena. In this regard I subscribe to the philosophy of logical empiricism. But I also believe that we must use these methods with caution, even maintain a degree of cynicism or distrust of them. Otherwise, we become vulnerable to the erroneous thinking that things that are not quantifiable are not meaningful or relevant. I see that insidious attitude all too often among economists, and it concerns me greatly. Our quantitative methods, in my view, are *means*, not ends, the same as our theories. I also offer the opinion that empirically quantitative research is not necessarily more sophisticated or useful than descriptive or interpretative research.

My interests and experience in economics as both a theoretical and an applied discipline undoubtedly affect my perspective and the tone of the book. In my opinion, those who belong to one group or the other both overemphasize the differences and dwell too much on them. Both groups have much to learn from one another, but the "territorialism" exists and is counterproductive. Some "theoretical" economists are somewhat prone to consider applied economics somehow inferior because it is applied (useful, directed at "mundane" contemporary issues, focused on problem solving). Some "applied" economists are somewhat prone to view theoretical or disciplinary economics as inferior because it is "irrelevant" (abstract, not recognized as useful by the general public). Disciplinary economics has an advantage in its theoretical sophistication, and applied economics has an advantage in applications of empirical techniques and making economics more obviously relevant. But we obtain good *applied* research only when we combine relevant elements of both in a single effort. Squabbling between the two "camps" reflects badly on both.

As for the subject of research methodology in economics, I find many of the writings on science, research, the "scientific method," and related topics as they relate to the research process to be by applied economists, while the writings on the same topics from the standpoint of economic methodology tend to be dominated by disciplinary economists. There are notable exceptions to this generalization, two of them being Glenn Johnson and Donald McCloskey, both of whom are cited frequently in this book. The distinction that I attempt to make between research methodology as it applies to the broad range of economic research activities and the more general, and philosophical, approach to economic methodology (economic reasoning) is in many cases a subtle one, but I believe it can be useful. I have references to selected material on economic methodology, but it is not my intent to deal specifically with that subject matter.

A final point on my views about economics, research, and methodology: the environment for research is critical. Research requires an atmosphere of questioning, seeking, and exploring without constraints on what is discovered. Many do not seem to grasp this obvious point. Analysis that has a goal of obtaining a preconceived result is not, by definition, research. Occasionally "research" is conceived as a means of supporting a political position or a business position or objective. Whatever the results of those efforts, they cannot be trusted because the approach was faulty. An equally insidious misuse is the squelching of research results when they do not confirm a preconceived result. I am not saying that this pitfall is unique to any particular type of institution (government, universities, corporations, etc.), merely that it is a reason to be wary of such research results. I recall a remark that administrators and bureaucrats cannot create good research, but they can surely prevent it.

ORGANIZATION OF THE BOOK

The remainder of the book is divided into three parts: the definitional, conceptual, and philosophical aspects of research methodology (chapters 2 through 4), the procedural aspects of research methodology (chapters 5 through 9), and the reporting of the research (chapter 10). Chapters 2 through 4 define terms and concepts, examine the methodology of the process, and consider how research and methodology are related to science, knowledge, objectivity, prediction, induction and deduction, and different philosophical beliefs. These chapters attempt to provide understanding, rationalization, and justification of the process as it is presented in subsequent chapters.

Chapters 5 through 9 are devoted primarily to planning and organizing a research project. We start with an overview of the research plan, which in written form is the research project proposal (chapter 5), then proceed to chapters organized around major elements of the research proposal. Some substantive issues on things such as suggestions and perspectives on writing, seeking funding support for research, and data collecting are included either as an adjunct to some chapters or as appendices. Chapter 10 deals with reporting the finished research.

A list of suggested reading has been included at the end of each chapter. These recommended sources, while not the only relevant sources of information on the topics covered in the chapters, augment the book or may be useful or thought provoking for readers in their own exploration of the subject. Some recommended exercises have also been included at the end of the chapters. These are provided as devices to help students in their efforts to integrate the concepts and guidelines into their own thought processes and approaches.

The book does not necessarily have to be covered in the same order of the progression of chapters, although it is probably best read in numerical order. For example, it may be beneficial to students to cover the material in chapters 1 and 2, then chapters 5 through 9, then chapters 3 and 4, with chapter 10 last. This sequence is especially helpful when, for example, students are required to develop a research proposal as a requirement of a course. Covering the procedural aspects earlier helps them get started on the proposal earlier in the term of the course.

Within the material of chapters 5 through 9, the order may be altered to, for example, chapters 5, 7, 8, 6, and 9. This approach might serve to emphasize that the problem definition, objectives, literature review, and the conceptual framework must, in the final analysis, be developed somewhat simultaneously rather than as a direct sequential progression. The guidelines offered are reliable, but the process is much too complex and relies too much on creativity to be reduced to a cookbook recipe approach.

SUMMARY

There is a need for formal, focused attention within the economics discipline on how to organize and conduct research in economics. This need has been acknowledged by the American Economic Association's Commission on Graduate Education in Economics as well as by individuals. This book has two central objectives: to provide (1) an overview of the conceptual and philosophical basis of research methodology in economics; and (2) procedural guidelines on planning, designing, and conducting research projects, especially applied economic research projects. The emphasis is on *research* methodology as it applies to economics, not economic methodology, philosophy of science, or research tools and techniques.

Economics and economic research topics are very diverse. Understanding, planning, and conducting research cannot be reduced to a checklist of steps or a universal formula. There are, however, guidelines and approaches that are time tested and reliable. Preparation to study research methodology requires disciplinary preparation and interest. It is also enhanced by a well-developed capacity for critical thinking, curiosity, synthesis, and emotional maturity.

The conceptual and philosophical aspects of research methodology are dealt with first in the book, then the procedural aspects. Some of the author's perspectives and biases are provided so that readers can approach the subject with added insight.

RECOMMENDED READING

Colander. "The Sounds of Silence."
Hansen. "The Education and Training of Economics Doctorates."
Krueger et al. "Report of the Commission on Graduate Education in Economics."
McCloskey. "Agon and Ag Ec."

NOTES

1. George Ladd was the first to present this perspective to the author.
2. This perspective was initially presented by George Ladd.

2

Research and Methodology

Wisdom is not a product of schooling but of the life-long attempt to acquire it.

Albert Einstein

The term *research* evokes images of activities that to many are often incomplete and in some cases misinformed. This happens with many specialized activities. Many people envision someone engaged in a laboratory, surrounded by test tubes, compounds, or objects being studied, or in a library, hidden by volumes of documents piled around the researcher. While both activities represent valid research functions, neither represents a comprehensive perspective of the research process, particularly in economics. The first may represent some realistic impression of laboratory research in chemistry or biology. The second may provide a valid impression of research activity in literature or history.

This chapter establishes some basic definitions and distinctions among some terms that are central to the study of research methodology in economics. It also provides some classifications of research and a characterization of the research process. The chapter is divided into five sections. The first section defines and discusses research. The second provides some classifications of types of research. The third defines and discusses methodology, making a distinction between methodology and methods. The fourth presents a description of the research process, and the last section emphasizes creativity and its role in the process.

RESEARCH DEFINED AND DESCRIBED

Many authors have written about research. Definitions provided for research, while not contradictory with one another, differ substantially. *Webster's Collegiate Dictionary* (1977) defines research as "studious inquiry or examination; *esp.*: investigation or experimentation aimed at the discovery and interpretation of facts, revision of accepted theories or laws in the light of new facts, or practical application of such new or revised theories or laws." Andrew and Hildebrand

15

(1982, p. 3) define research as "the orderly procedure by which man increases his knowledge." Ghebremedhin and Tweeten (1994, p. 4) define it as applying "the scientific method to study hypothetical propositions of presumed relations among phenomena." Leedy (1985, p. 4) says research is "the manner in which we attempt to solve problems in a systematic effort to push back the frontiers of human ignorance or to confirm the validity of the solutions to problems others have presumably resolved." In many cases (e.g., Committee on Science, Engineering, and Public Policy, 1995), authors discuss and attempt to describe research without trying to define it.

While each of these definitions makes a meaningful statement about research, the following is offered as an accurate and comprehensive, yet succinct, definition:

> **Research is the systematic approach to obtaining and confirming new and reliable knowledge.**

This definition identifies several characteristics that are essential for an accurate and comprehensive definition. First, research is not limited to certain types of activities such as laboratory methods or literature searches. Additionally, the definition indicates that research is systematic and orderly.[1] As an approach, research follows a sequence or a series of steps (with variations in the components and their order). Lastly, its purpose is new knowledge (what is not already known), and the knowledge must be reliable (demonstrable to others based on reason or evidence). The emphasis is on both *new* and *reliable*, although both are not always dealt with in the same individual effort. For example, some research may focus entirely on establishing reliability (i.e., *confirming*) of things we think we know.

While our specific interest is research in economics and thus we are interested in a definition of research that is workable for economics, the above definition is applicable irrespective of the discipline. We must, however, understand the similarities and differences in how the research process is conducted in economics and other disciplines. Even within economics, research can be done in many different ways. It is also useful to understand how other disciplines may validly differ, if only to keep ourselves from becoming inflexible in recognizing avenues to reliable knowledge. Therefore, both differences and similarities with other disciplines will be periodically acknowledged.

Note that the term *truth* was not used in the definition of research. Some authors (e.g., Larrabee, 1964) use the word, but we will avoid it within the context of this book. Truth implies something known with certainty, something beyond questioning. This concept of truth is outside the productive realm of thinking by researchers. Truth is a luxury we may allow ourselves in our private beliefs, but is a potentially deadly attitude in research. To "accept" an outcome as either impossible or certain is to deny the possibility of that outcome being different from what we think we already know. The statement attributed to Will Rogers, a twentieth-century American humorist and philosopher, reflects the spirit of this attitude: "It ain't what we don't know that gets us in trouble; it's what we know that ain't so." Or, in the words of Nobel physicist Richard Feynman (1999, p. 248):

> To make progress in understanding, we must remain modest and allow that
> we do not know. Nothing is certain or proved beyond all doubt. . . . And as
> you develop more information . . . , it is not that you are finding out the truth,
> but that you are finding out that this or that is more or less likely.

When we use *truth* in a research setting, it is with a limited meaning. This will be addressed in more detail later in the book.

When describing research, it may be useful to specify what it is not as well as what it is. Research is ***not***:

A. *Accidental discovery*, although new knowledge may result from accidental discovery, and accidental discovery may occur within a structured research process. Accidental discovery, when it occurs, usually takes the form of a phenomenon that has not been previously noticed. Such a discovery or observation *may* lead to a structured research process that is directed toward verifying or understanding the observation. That process may in turn lead to new reliable knowledge.

B. *Data collection*, although gathering data is frequently a part of research. The data themselves may not constitute, in the economist's perspective, "reliable knowledge." The data are an intermediate step toward knowledge, which entail relationships among forces, factors, or variables. We should acknowledge that gathering data in some fields (e.g., applied statistics) may constitute the research process; however, gathering data so that it constitutes a statistically valid sample of a population is an objective. Other disciplines, including the laboratory sciences (chemistry, biology, etc.) and some field sciences that attempt to apply the controlled experimental design to situations without all conditions controlled may be concerned primarily with producing valid experimental data, which are used to test a hypothesis, typically of a simple causal relationship. To the economist, those data may provide the means to achieve another objective. Examples include the examination and evaluation of more complex relationships and testing hypotheses about them.

C. *Searching out published research results in libraries*, although this is an important part of the research process. The point is that the literature search in economics is an important early step, but research does not stop there. The research process always includes synthesis and analysis, even if it is directed at assessment of the prior studies. Some disciplines (e.g., literature and history) rely more heavily on synthesis and analysis of prior studies as the major component of the research process than does economics. The general public's perception that research consists merely of reviewing the literature is misleading because that process, by itself, may produce no new knowledge. Producing new knowledge from prior literature alone must involve, at a minimum, synthesis of the prior material.

Research is:

   A.   *Searching for explanations* of events, phenomena, relationships, and causes, at least in economic research. Some research is occupied primarily with estimation or determination of relatively simple (e.g., two-variable) relationships—*what* occurs when something else happens. We usually want to know, in addition, *how* it occurs or *why* it occurs. Other research addresses more complex relationships, often dealing with interactions among the simpler relationships. Both are important. We often must grasp the simpler relationships before progressing to the more complex ones.
   B.   *A process.* As such, research is also planned and managed. Unless structured carefully, the research knowledge cannot be reliable (demonstrable to others on the basis of generally accepted rules of evidence). As a process, it also is usually interactive. Research is not done in isolation from prior research or stimulation from peers and coworkers. The process is also creative. From one perspective, we are creating new understanding. From another perspective, each research activity (project) is in some respects unique and requires a design specific to that activity. Research also is circular—it always leads to more research questions.

   A distinction between research as it relates to physical or biological phenomena and research as it relates to social phenomena, of which economic research is a part, is useful. Because the nature of what is being studied in each kind of research has different characteristics with different capacities for direct observability/measurability, each demands some variations in approach to extending human knowledge about them (Committee on Science, Engineering, and Public Policy, 1995, p. 3). In addition, there seems to be something vague and mysterious about economic research. Perhaps most people intuitively know that they do not grasp what the activity includes. This contrasts with a public perception, correct or not, that people have a realistic grasp of the role and functioning of other research fields such as medical research or engineering research. Embodied within this attitude may be the notion that economic research is an activity or exercise confined to academia and a few government agencies or foundations. Even more damaging is the perception that economic research is without practical application: that is, it is not "useful."
   We maintain that these perceptions are without basis. *All* well-designed and conducted economic research has potential application. The applications may not be immediately obvious in highly theoretical, "basic" research and may be more obvious in more "applied" research efforts. One must realize, however, that the applied research is usually dependent on certain basic research being or having

already been done, so that the basic research capital stock must be maintained to produce the applied results. Also, the two often progress hand in hand and are difficult to distinguish. Well-developed and executed research, whether basic or applied, focused on defined issues always has application toward relevant problems or issues by persons in the discipline or by public and/or private decision makers. The failure to see its applications has two primary sources: (1) some users of research cannot easily grasp the applications of research because they are not formally trained or experienced in the specialized methods of economic research and reasoning; and (2) researchers often do not provide adequate interpretations and guidance on application (use) of the research to the potential users of their research. It is the responsibility of researchers to help users of their research to understand its implications and to help ensure that the research is used effectively and not to draw invalid implications.

Research takes place in the private sector as well as the public sector. The greater proportion of *published* research is associated with researchers in universities and government, where widespread dissemination of research procedures, results, and conclusions is regarded as an important part of the institutional missions. Many business firms, industry organizations, and consulting firms, for example, likewise conduct research. There *may* be a tendency for research conducted in the public sector to be more rigorous and objective because it tends to be subject to greater scrutiny, but the contention is at best an untested hypothesis. This does not suggest that any research result is more or less reliable on the basis of its source. But "research" by a firm that evaluates the benefits of a product in which the firm has a financial interest may need to be examined with greater caution than an evaluation by a group with no stake in the outcome. Nevertheless, much research is conducted within the private sector, and the validity of that research is as important as that in the public sector.

CLASSIFICATIONS OF RESEARCH

The characterization of thought, actions, and processes into classifications or divisions of general types helps us make distinctions. Before we can conceptualize, we must first *define* and then *classify* (Piaget, 1954, p. 357). Definitions and classifications are addressed throughout this book because they are fundamental to understanding the subject matter. Their necessity is explained by Bloom et al. (1956). Knowledge of various levels and types fosters intellectual abilities or skills. The hierarchy of knowledge runs from knowledge of specifics (including terminology and facts), to knowledge of dealing with facts (including classifications, methods, and methodology), to knowledge of universals and abstractions in a field (including principles, generalizations, theories, and structures). Intellectual abilities, while not completely dependent on knowledge, influence one's mastery of knowledge. The spectrum of intellectual abilities and skills progresses from comprehension (translation, interpretation, and extrapolation), to application. It then progresses to analysis (of elements, relationships, and organizational principles), to synthesis (of communication, plans, or abstract relations), and to evaluation.

In classifying research, there are different criteria on which to base the classifications, and they may each be useful for specific purposes. A brief discussion of several of them is provided below. These classifications are introduced here, but several will be discussed again in subsequent chapters.

Basic vs. Applied Research

Definitions of these terms vary. *Basic research* can be characterized as that which attempts to determine or establish fundamental facts and relationships within a discipline or field of study. Andrew and Hildebrand (1982) define applied research as "research undertaken specifically for the purpose of obtaining information to help resolve a particular problem" (p. 3). Applied research has a definite purpose (p. x). A major distinction between basic and applied research is the focus of its application. Basic research is conducted with relatively little emphasis on its applications to "real-world" policy and management issues, while applied research is conducted with those issues at the forefront. This classification is of limited usefulness because it lacks clarity in its distinctions. For example, some basic research is conducted to develop a method, measurement, or theory so that an applied problem can be addressed. Is that research "basic" or "applied"? There is obviously more than one defensible position on the question.

The attempt to distinguish between basic and applied research may have its roots in attempts to classify basic and applied science. The premise of this distinction is that basic science ignores applications to current societal problems, while applied science focuses largely on those problems, sometimes without due concern for the underlying causal forces and relationships. This distinction is counterproductive to the extent that it causes scientists and researchers to limit their perspective of the relevance of their work. Nobel Prize recipient Herbert Simon (1979, p. 494) addressed the somewhat artificial distinction as follows:

> It is a vulgar fallacy to suppose that scientific inquiry cannot be fundamental if it threatens to become useful, or if it arises in response to problems posed by the everyday world. The real world, in fact, is perhaps the most fertile of sources of good research questions calling for basic scientific inquiry.

Disciplinary, Subject-Matter, and Problem-Solving Research

This classification of types of research as described by Johnson (1986b) is perhaps the most comprehensive and useful one developed to date; reading of his treatment of the classification is important for the serious student of methodology. Breimyer (1991, p. 247) likewise recognizes the prominence of Johnson's classification contribution to the study of research methodology. Figure 2.1 provides a characterization of the overlap between the basic/applied research classification and Johnson's disciplinary/subject-matter/problem-solving research classification.

By way of general description, *disciplinary research* is that which is "designed to improve a discipline" (Johnson, 1986b, p. 12). It dwells on theories, fundamental relationships, and analytical procedures and techniques within the discipline. Disciplinary research in economics is the research for which the intended users are other economists. While purely disciplinary research is not the

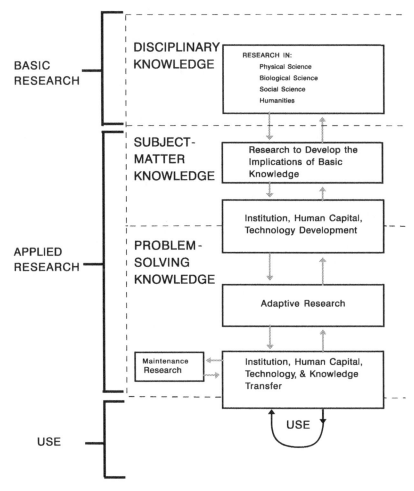

Figure 2.1. Relationship among research, knowledge, and use (Bonnen, 1986).

total research activity within economics, it is of utmost importance because it provides the conceptual and analytical base for the other types of economic research. Consequently, disciplinary research is synergistic and complementary with subject-matter and problem-solving (multidisciplinary) research; disciplinary research provides the foundations for much of the more applied research, while the applied research often reveals disciplinary shortcomings that need further disciplinary research (Johnson, 1991, p. 188).

Subject-matter research is "research on a subject of interest to a set of decision makers facing a set of practical problems" (Johnson, 1986b, p. 12). This type of research tends to follow subject-matter boundaries within a discipline—for example, resource economics, production economics, labor economics—and is inherently multidisciplinary because it draws information and ideas from other disciplines. Subject-matter research provides policy makers with the concepts and knowledge with which to make decisions about the sets of problems they address. Subject-matter research in economics commonly involves integrating subject-matter or issue knowledge from other disciplines such as engineering, sociology, or chemistry, with disciplinary (economic) concepts to draw implications about probable effects of proposed or possible actions. The distinctions between disciplinary and subject-matter research can be illustrated with the following examples. Medicine, a subject-matter field, utilizes the disciplines of biology, psychology, and chemistry (as well as other disciplines); management uses the disciplines of economics and psychology; engineering uses physics, mathematics, chemistry, and other disciplines. Subject-matter research is a cornerstone in economics in that it represents the direct application of economics to contemporary issues. Many economists consider subject-matter research to be directly relevant to current social and public issues and to be the primary source of policy applications for the discipline.

Other disciplines may be becoming more focused on subject matter. For example, biochemistry has adapted or integrated biological concepts and knowledge with chemistry concepts and knowledge, driven by a concern to develop useful, multidisciplinary knowledge. Bonnen (1986) goes further to suggest that this increased multidisciplinary focus is driven by public dissatisfaction with disciplinary research of unknown relevance (i.e., without specific purpose) and the corresponding shifts in research funding. He states that "to practice at the cutting edge in almost any discipline today, even in the biological and physical sciences, requires not only command of a discipline but also of major components of knowledge from related disciplines well beyond mathematics and statistics" (p. 1073). These types of subject-matter–driven integrations sometimes evolve into new disciplines. We may be seeing biochemistry evolving toward a distinct disciplinary status, even though its focus is largely subject-matter–oriented. All engineering fields (civil, electrical, industrial, petroleum, textile, etc.) are subject-matter–oriented and multidisciplinary. They draw from the theoretical realms of the "basic" sciences and mathematics and address problems defined in part by external (often industry) parties. Subject-matter areas in colleges of agriculture (e.g., agronomy, animal science, horticulture) depend on their "basic" parent disciplines (botany, zoology, chemistry, etc.).

Economics is a separate discipline because it has its separate, distinct body of theory and empirical knowledge. The subject-matter areas using economics are inherently multidisciplinary. For example, consumer economics draws from psychology, natural resource economics from biology, and economic policy from political science. The various subject-matter areas are dependent on a common disciplinary base.[2] Thus, while economics is a separate discipline, much of what we eventually do with it—its applications—becomes multidisciplinary subject-matter work.

Problem-solving research is designed to solve a specific problem for a specific decision maker (Johnson, 1986b, p. 13). As is subject-matter research, it is inherently multidisciplinary because the problems that individual decision makers face invariably span disciplinary interests. Problem-solving research has a heavy focus on prescriptive knowledge (and it uses both positivistic knowledge and knowledge of values, which will be defined and discussed in chapter 3). Problem-solving research is more directly linked to a particular decision process than is subject-matter research. Problem-solving research is not more problem-oriented than the other types (all research is directed to a defined problem), but the problems it addresses are more oriented to specific, individual applications.

Problem-solving research often results in prescriptions or recommendations on decisions or actions. There is a difference of opinion on the delineation between problem-solving *research* and other problem-solving activities. Some view the problem-solving research activity as extending into the researcher being in the decision and implementation phases of the problem-solving process. The perspective advocated here is that the problem-solving research process stops short of implementation involvement. However, the distinction may not be critical in most respects. While the researcher may not make the decision or implement it, the prescriptions or recommendations become close to those activities.

Problem-solving research is holistic—it uses all information that is relevant to solving the specific problem at hand—while disciplinary research is reductionist (Johnson, 1987, p. 88), focusing only on the disciplinary content of the problem, often reducing the other components to a set of assumptions. Subject-matter research reaches across disciplines but is not focused on specific decision issues, and is therefore less holistic than problem-solving research. Disciplinary research is generally the most durable because the fundamental knowledge tends to maintain itself over a long period, while problem-solving research is generally the least durable because the situations addressed are specific and the least generalizable (Johnson, 1986b, ch. 16). Johnson (1986b) also notes that the three types of research must be judged by different criteria and perhaps from different philosophical perspectives. Johnson (1997, p. 261) states the case as follows:

> When the object is to generate disciplinary knowledge, rigor can be regarded
> as disciplined careful efforts to improve the theoretical and empirical knowl-
> edge and techniques of an unambiguously defined discipline. . . . Multidisci-
> plinary problem-solving work is often appraised using standards more appro-
> priate for disciplinary work . . . (and) often foolishly denigrated . . . despite
> objectivity and care in carrying it out.

One way of synthesizing the fundamental differences, and some of the conflicts, between disciplinary and multidisciplinary perspectives with this classification system for research is as follows:

> Those with responsibility for solving a problem or resolving an issue tend to
> view disciplinarians as potential consultants or advisors. Disciplinary knowl-
> edge becomes something to put in a multidisciplinary information retrieval
> system. . . . Until disciplinarians learn to suppress their disciplinary chauvin-
> ism, they are poorly qualified to administer problem-solving and issue-
> oriented investigations or to be decision makers. Disciplinarians and applied
> disciplinary knowledge by themselves typically cannot solve a problem or
> resolve an issue because of their specialized natures and because disciplines
> neglect interdependencies among the kinds of knowledge generated in differ-
> ent disciplines that are crucial in addressing the problem or issue at hand.
> (Johnson, 1997, pp. 258–259)

Analytical vs. Descriptive Research

Descriptive research may be characterized as simply the attempt to determine,
describe, or identify what is, while *analytical research* attempts to establish why
it is that way or how it came to be. The intent of descriptive research is often synthesis rather than analysis. Synthesis, or description, attempts to pull knowledge
or information together and decipher its logically plausible connections (synthesize) rather than take the information apart to examine why and how it came to be
(analyze). All disciplines generally engage in both types, although there may be
differences in relative emphasis among disciplines. An example of descriptive
research in economics is that which establishes the structure of an industry—
number, size, location, and so on of firms within an industry. The corresponding
analytical research on that same topic might be to determine the behavior of that
industry in terms of pricing and other competitive behavior.

Economists see their discipline as having a pronounced analytical focus.
Even so, we make extensive use of descriptive analysis in our research. We often
employ descriptive analysis in the process of understanding, explaining, and documenting the problems on which we conduct research. We typically do this by
gathering information and data, then putting them together in a way that it "tells a
story" about the research problem or issue. An example is using data to show how
real income of a group has changed over time, possibly in relation to the cost of

living for that same group. The data may be used to demonstrate a discrepancy between their movements and justify the need for more information on why the incomes and costs have moved as they have. Another way that descriptive research is widely used in economics is in subject-matter textbooks. While these textbooks may report results of analytical research, they tend to be based largely on making sense of the mass of information on the subject.

METHODOLOGY DEFINED AND DESCRIBED

The definition of methodology and the differentiation between methodology and method are matters on which reasonable people may disagree. The two terms are used interchangeably in many contexts. Rather than engage in a debate, methodology and methods are defined and delineated as they are used in this book.

Authorities are somewhat consistent in their definitions of methodology. Runes (1983, p. 212), in his *Dictionary of Philosophy*, defines methodology as "The systematic analysis and organization of the rational and experimental principles and processes which must guide a scientific inquiry. Also called scientific method. Thus, methodology is a generic term exemplified in the specific method of each science." *Webster's Collegiate Dictionary* (1977) defines methodology as "a body of methods, rules, and postulates employed by a discipline; . . . the analysis of the principles or procedures of inquiry in a particular field." Similarly, *Webster's New World Dictionary* (1968) defines it as "the science of method, or orderly arrangement; specifically the branch of logic concerned with the application of the principles of reasoning to scientific and philosophical inquiry . . . a system of method, as in any particular science."

Definitions of method are considerably more varied. *Webster's* (1977) definitions of method include "a way, technique, or process of or for doing something," and *Webster's* (1968) definitions include "a way of doing anything; mode; procedure; process; especially, a regular, orderly, definite procedure or way of teaching, investigating, etc." Breimyer (1991, p. 248) maintains that the term *method* is ill defined.

Using these definitions, the term *methodology* in this book refers to *the study of the general approach to inquiry in a given field*. Thus, research methodology in economics is the study of the general approach to research in economics. The term *method* is used to refer to *specific techniques, tools, or procedures applied to achieve a given objective*. This use of the term is consistent with that of the National Academy of Sciences' Committee on the Conduct of Science, which stated:

> Throughout the history of science, some philosophers and scientists have sought to describe a single systematic method that can be used to generate scientific knowledge. Some scientists may believe in such a picture of themselves and their work, but carrying this approach into practice is impossible.

> Rather than following a single scientific method, scientists use a body of
> methods particular to their work. . . . these methods include all of the tech-
> niques and principles that scientists apply in their work and in their dealings
> with other scientists. (Ayala et al., 1989, p. 2)

Research methods in economics include various versions of regression
analysis, mathematical analysis, operations research techniques, surveys, gathering data, use of selected theoretical constructions, and other procedures,
including combinations of techniques. Using this classification, *methods* and
methodology are not interchangeable; methods are a portion of the concern of
methodology.

These definitions may encounter some objections from some authors of journal articles, particularly in recent years. It has become commonplace to use the
term methodology to describe, for example, specific analytical techniques such as
Markov chains or certain econometric methods. However, if we are to apply rigor
to the meanings of terms as we apply them in the study of research, definitions
such as those above are necessary. Paarlberg (1963, p. 1386) offers the view that
many economists "use the word methodology when they are concerned neither
with philosophy nor with logic, but simply with method. The three syllables are
added to convey prestige."

This point is also addressed by Machlup (1978). He states that "language
pollution . . . has led to a vulgar use of the word methodology in a sense violating
all philosophical traditions." He attributes this pollution to people being unin-
formed about meanings and too lazy to be precise; in this case adopting methodology as a synonym for method. Machlup states:

> The propensity of many people to substitute technical terms for ordinary,
> well-understood words from the vocabulary of everyday speech has long
> been an effective force in the development of language, often leading to its
> enrichment, but sometimes to its pollution. The adoption of the term *method-
> ology* by people ignorant of its original and proper meaning is a case of lan-
> guage pollution, probably of the irreparable sort, because we now find the
> word more often misapplied than not. (p. 6)

We also should make a clear distinction between research methodology in
economics and economic methodology. *Research methodology in economics* is
the study of the approach to research in the discipline of economics. *Economic
methodology* is the study of the general approach to economic reasoning and validation of economic concepts. Economic methodology relates to the disciplinary
basis (reasoning) of economics. Research methodology focuses on the process of
developing information and knowledge, which may provide application knowledge and/or additional disciplinary understanding.

Boland (1982, p. 1) defines economic methodology as the view of the relationship between economic theories and methods of reaching conclusions about

the nature of the real world. Hausman (1989, p. 115) states that "economic methodology is concerned mainly with questions of theory confirmation or disconfirmation or empirical theory choice." Blaug (1980, p. xi) presents economic methodology as "philosophy of science applied to economics." The explanations of economists consist of theories. Economic methodology deals with questions of

> what is the structure of these theories, and in particular, what is the relationship between assumptions and predictive implications in economic theories? If economists validate their theories by invoking factual evidence, is that evidence pertinent only to the predictive implications of theories, or to the assumptions of theories, or both? (Blaug, 1980, p. xii)

Other useful references on economic methodology include Caldwell (1982) and Stewart (1979).

While economic methodology and research methodology in economics are different, they are not independent. They depend on one another in much the same way as do science and research, which are discussed in chapter 3. Economic reasoning or logic and the processes by which we pursue economic knowledge is of necessity an integral part of the economic research process. Conversely, the ways we conduct research affects the body of accumulated economic knowledge and concepts. Yet if we do not mentally separate them, we risk the error of mistaking, for example, the process of selecting an appropriate model or concept for the process of evaluating whether the model or concept applies. Taking the position that a theory is applicable in a given situation and testing its applicability in that same situation are *not* the same; the former is likely a prelude to or a part of the latter.

In some respects, research methodology might be viewed as a subset of economic methodology because the research process is a means for developing and evaluating economic theory. However, that perspective may be misleading. The two methodologies are more accurately viewed as symbiotic—each depends on, and supports, the other.

This symbiotic relationship can be illustrated by drawing from research on the general topic of hedonic or characteristic price analysis. The theoretical basis for hedonic/characteristic price analysis lies with the works of Lancaster (1966) and Rosen (1974), with several notable prior and subsequent works. Research methodology is employed in conducting specific and general exploration of the valuation of characteristics or attributes embodied in goods. Research methodology is concerned with how that analytical process is undertaken—developing models (based on theory, which is, in turn, related to economic methodology) for analyzing the relationships and parameters, selecting and obtaining data with which to estimate relationships and test hypotheses, evaluating implications of the relationships, and so on. What is learned, in turn, may provide additional insight into how markets work to value attributes or characteristics: that is, how we *think about* markets and their operation (our theories about markets). The new insight, understanding, and reasoning (our slightly expanded and/or altered theoretical base) then provide an expanded base of theory on which additional research can be built.

THE PROCESS OF RESEARCH

As indicated earlier, research involves or constitutes a process. While the specifics of the way in which the process may be implemented or applied has infinite variations, figure 2.2 is a stylized representation of the process. The process is initiated with a question or problem—the researcher's perceived need for knowledge about something. The question or problem can come from various sources, and we can make some generalizations about the types of research that are more likely to ensue depending on the source(s) of the problem(s). As with all generalizations, they are fallible. Questions exclusively within the discipline are likely to be theoretical or procedural questions that lead to disciplinary research. Questions coming in total or in part from related disciplines, the government, or public-interest groups are more likely to lead to subject-matter or problem-solving research. Questions influenced by firms or industry groups are apt to be problem solving in their orientation.

Given the question, problem, or issue, the next step is to formulate goals or objectives to deal with the problem. These objectives, as with the problem, are

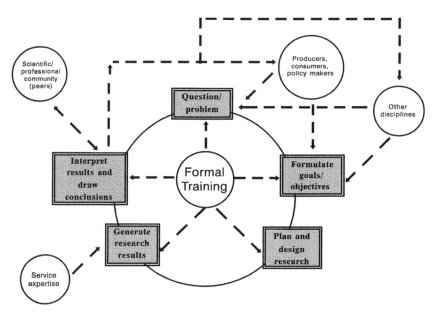

Figure 2.2.   Schematic of research process.

likely to be influenced by the research clientele—other economists, industry, the public, or other disciplines. As part of a *research* process, these research goals relate to obtaining knowledge, information, or understanding about the problem/issue/question. Research objectives or goals may include deriving prescriptions and/or recommendations, especially when problem-solving research is involved.

The third step in the process, research design, becomes more isolated from other parties when it involves specialized techniques. It is, however, influenced by solicited specialized expertise from within or outside the discipline. For example, we might seek theoretical or modeling insight from colleagues, such as mathematicians or statisticians, who are more experienced or specialize in certain topics. The research design is formulated to achieve the research objectives, may contain many elements, and is the basis for accepting research findings. The reliability of results from a given research endeavor depends on the validity of the research design, never on its findings.

The fourth step, that of generating or producing results, is often a more "mechanical" step in comparison to the steps that precede it, especially in empirically oriented economic research. For example, this may be the step in which we generate parameter estimates by using statistical techniques from econometric model(s) formulated in the research design, or make projections using mathematical simulation model(s) constructed in the research design. In empirical research, service expertise might include computer-programming input. If the research is not empirical in nature, it is the implementation of whatever research plan was formulated in the previous step and is likely not "mechanical." In theoretical disciplinary research, this step may consist largely of the application of deductive reasoning to the premises and situation being analyzed.

Once results are obtained, they must be interpreted and analyzed themselves (step 5). This is the step in which we look at implications, draw conclusions, and perhaps offer prescriptions. For example, empirical results from an econometric model are analyzed and interpreted to determine what they show in terms of policy and management applications. We may derive estimates of demand and supply elasticities from an econometric market model, then examine the implications of the elasticities and compare the elasticity estimates to those estimated from other studies. Step 5 is the one most often overlooked or underestimated by students as they are being initiated to the research process. They often see the empirical results as the end product of the process. Actually, the interpretation of those results may be more challenging, and more important, than the process of obtaining the empirical estimates.

Once step 5 is completed, results are typically disseminated. Dissemination within the discipline occurs through disciplinary and subject-matter journals, professional meeting presentations, and other types of research reports. Dissemination to the public occurs through research reports, presentations, and occasionally public media coverage. Dissemination to related disciplines occurs through research reports, presentations, and multidisciplinary journal papers. Dissemination within a corporation or government agency may consist of an internal research report and/or an oral report. Presentation to the different audiences must vary because of differences in background and expertise of the clientele. The different types of audiences are each important in their own way and maybe for different reasons.

The solid line between step 5 and step 1 (figure 2.2) closes the loop and makes the research process circular. Getting answers to some questions always leads to other questions. The more we understand about a given phenomenon, the more clearly we see the particular aspects of it that we do not yet understand. The ensuing questions may be in the same vein as the research questions originally posed, or they may be more peripheral to the original question. It is this characteristic, however, that evidences the need for the process to be planned and organized. Unless we impose some limits or bounds on an individual research effort, the tendency is for it to be a continuous question-answer-question progression. By imposing some closure on an individual effort, the process becomes one of a succession of research projects, each potentially linking to others and the accumulation of knowledge about the topic. While slower than we may prefer, the process is at least systematic. By periodically disseminating procedures, results, and conclusions, information is released to others working on the same or similar topics. The subsequent feedback from extended research users, represented by the dotted lines between the public policy makers (figure 2.2), is an important part of the loop, especially with problem-solving and subject-matter research. Users react to research output, then provide additional questions and issues for research.

CREATIVITY IN THE RESEARCH PROCESS

Research is a creative process. This is a point that cannot be overemphasized. Brewster (1970, pp. 234–235) states that

> there is much to the research quest that cannot be reduced to systematic rules. . . . research includes far more than mere logic, whether deductive or inductive. It includes insight, genius, groping, pondering—"sense" which can't be boxed up in any formalized procedures. The logic we can teach; the art we cannot.

The research process requires, or at least works best with, imagination, initiative, a degree of intuition, and a large dose of curiosity. There is no "magic formula" for accomplishing or producing good economic research. If it were that simple, anyone could do it. There may be, and often are, different ways to accomplish a given research mission or objective. We may occasionally even conduct research to determine which approach is more effective or efficient.

Creativity is a somewhat vague concept on which written material seems to provide no clear definition. Creativity seems somewhat like "art"—we cannot define it, but we know it when we see it. For our purposes, it is defined as the generation and implementation of new ideas. One reason for confusion about creativity is that there are different types of creativity (Stuller, 1982, pp. 38, 39):

   A.   *Imaginative creativity.* This is creativity through imagination— purely imaginary (non-factual) mental constructions. This type of creativity is usually associated with painters and writers. However, economists engage in imaginative creativity when we, e.g., formulate hypothetical relationships or causal linkages which we then proceed to evaluate and test.

   B.   *Natural creativity.* This is creativity through natural physical or mental ability and is usually exhibited by performance. This type of creativity is associated with dancers, musicians, and athletes. Natural creativity has little to do with economics unless it is extended to the sphere of mental activity. Perhaps some have a greater capacity for economic logic than others.

   C.   *Prescriptive creativity.* This is creativity through development of new possibilities. It is stimulated by "what if . . ." questions, and tends to be associated with philosophy and religion. Economists engage in prescriptive creativity in, e.g., arriving at public policy prescriptions and guidelines.

   D.   *Applied creativity.* This is creativity in the form of application of concepts to solve problems. It is not associated with any field in particular, but is an activity in any "problem solving" discipline. Applied creativity is an integral part of economists' activities because our activities are oriented largely to addressing societal and managerial problems.

   E.   *Theoretical creativity.* This is creativity through the development of relationships between things; it results in abstract concepts. Theoretical creativity is a part of most academic disciplines, although not necessarily all fields of endeavor. Economists draw on theoretical creativity in their disciplinary theoretical work.

The degree to which applied and theoretical creativity can be separated, at least in economics, may be questioned. For example, Bonnen (1986, p. 1072) presents the view that "Much of the creativity in any discipline comes from the intellectual stimulation of confronting disciplinary knowledge with the test of application."

Intuition and imagination may be important parts of all types of creativity. Creativity is rarely achieved without tenacity and perseverance, and the evidence of creativity is not often realized without work. Psychologists believe that creativity is an acquired ability to a large extent—it can be learned (Gilbert, 1986), which would appear to be consistent with the view that creativity can be acquired. The American Economic Association's Commission on Graduate Education in Economics expressed concern that graduate programs may be stifling creativity by overemphasizing acquisition of technical skills in the early years of graduate programs. The commission recognized that "creativity relevant to economic research is multidimensional. It can be manifested by posing a new question, by reformulating an existing problem, by finding new data sources with which to test a proposition, by developing a new model, or by extending mathematical or statistical techniques" (Krueger et al., 1991, p. 1049).

We lack prescriptions on how to be creative, much the same as there are no prescriptions for how to be "talented" or how to be an effective communicator. There are, however, some things you can practice that will foster creativity (as there are certain guidelines that will help you become an effective communicator). Stuller (1982) provides a set of suggestions for fostering creativity, and Ladd (1987) identifies conditions conducive to creativity. These provide the basis for the following guidelines for helping you to be creative:

A.   Gather and use previously developed knowledge; thorough preparation stimulates unconscious thought processes.

B.   Exchange ideas with other people, especially your colleagues.

C.   Capture your intuition by applying deductive logic. When your intuition suggests something, make note of it before it escapes, then deal with it on a logical, analytical level.

D.   Cultivate the ability to look at things in alternative ways. Diversity of interests and experiences, as well as maintaining a degree of doubt about almost everything, helps to develop this ability.

E.   Question or challenge assumptions.

F.   Practice searching for patterns or relationships among occurrences or things. This includes Ladd's sensitivity to similarities (perception of a link between things not previously connected) and using combinations of the above practices.

G.   Do not be afraid to take risks; little can be gained if you are not willing to explore. Ladd refers to this as developing a venturesome attitude. A statement (source unknown) may illustrate the point: moderation is fine for parents and politicians, but science, literature, and the arts need boldness.

H.   Cultivate a degree of tolerance for uncertainty. When you know things that are contradictory, give yourself time to ponder them and seek a resolution; a temporary suspension of judgment may allow the unconscious mind to reconcile the contradiction.

I.   Allow your curiosity—your intense desire to know or understand—to build a creative tension.

J.   When you are unable to resolve a troublesome problem, try abandoning it, at least in your conscious mind, for a while. This may let the unconscious mind work on it and provide new insight when you return to it with your conscious mind later.

K. Discipline yourself to write down your thoughts. There is something about having to express your thoughts in writing, so that they are clear to someone else, that fosters your own process of mental discovery. Ladd (1987, p. 77) states, "Many of my intuitions come to me when I am writing. It frequently is true that I do not know what I think until I write it." More will be said about the role of writing in chapter 5.

L. Create situations in which you have freedom from distraction. Having some protected time in which to concentrate and contemplate may be necessary for creative thinking.

Psychology provides us with the understanding that the two hemispheres of the brain are somewhat specialized. The left side is more dominant in functions involving abstract reasoning and language, while the right side is more dominant in visual and perception skills, intuition, and expression of emotion (Levy, 1985; Buffington, 1986). The left hemisphere processes information sequentially, while the right hemisphere processes information simultaneously (Ladd, 1987, p. 29). Each side is capable of functioning independently of the other and they work differently and come up with different answers. Austin (1978, p. 139) explains that the left hemisphere "proceeds, piecemeal, to examine the irregular bark on each tree . . . [while the] right hemisphere grasps in one sweep the shape of the whole forest, relates it . . . to the contours of the near landscape, then to the line of the horizon." However, creative thinking involves a free flow of messages between conscious and unconscious (Ladd, 1987, p. 27); creativity is produced by integrating the abilities of both sides. The interaction creates a special mental procedure that exceeds the capacity of the independent parts.

This spark of creativity, if you can find it, may be the difference between satisfactory and outstanding research. Perhaps good (i.e., useful) research may be done without creativity, and creativity does not guarantee good research. However, the most innovative research probably requires creative thinking.

SUMMARY

Research entails different activities and approaches among disciplines, and the public's comprehension of economic research may arguably be more limited than it is of research in fields such as medicine, engineering, and chemistry. This chapter establishes some definitions of essential terms, classifications of types of research, distinctions between sometimes confusing terms, and an overview of the research process in economics. The importance and role of creativity is emphasized.

Research is defined as the systematic approach to obtaining new and reliable knowledge. Activities such as accidental discovery, data collection, and reviewing published research do not, by themselves, constitute research, but each plays a part in research. Economic research is a structured, interactive, creative, circular process of searching for explanations to economic problems and phenomena. Research is presented under three different classifications: (1) basic and applied research; (2) disciplinary, subject-matter, and problem-solving research; and (3) analytical and descriptive research. The different classifications are each useful.

Research methodology—the study of the general approach to the research process in economics—is distinguished from research methods—particular techniques, tools, or procedures used to meet a set of research objectives. Also, research methodology in economics is distinguished from economic methodology. The process of research methodology in economics is described as being circular with certain steps or components. The steps consist of problem/question/issue specification, objective identification, design of methods and procedures, generation of results, and interpretation of results. The process typically identifies a new set of research problems.

The role of individual creativity and imagination in the process is emphasized, with clarification and examples provided on different types of creativity and how each may be relevant in economic research. Suggestions on how to foster creativity are also provided.

RECOMMENDED READING

Bloom et al. *Taxonomy of Educational Objectives*, pp. 201–207.
Hausman. "Economic Methodology in a Nutshell."
Johnson. *Research Methodology for Economists*, pp. 1–7, 11–14.
Ladd. *Imagination in Research*. The entire book is recommended reading, but most especially parts I and II, pp. 1–94.
Machlup. *Methodology of Economics*, pp. 5–10.

SUGGESTED EXERCISES

1. Describe an example of each type of creativity identified in this chapter from your own experience.

2. Select three articles, each from a different recent issue of an economics journal. Classify each article according to the type(s) of research reported in the article and explain why you classify them as such.

3. Select three recent journal articles (they may be the same as in exercise 2 above). Is the term *methodology* used in each? If it is, is the term used as it has been defined here or is it used to describe a method or technique?

4. Of the techniques described to help foster creativity, which do you use regularly? Describe, in writing, how you employ them.

NOTES

1. This does not say that the outcomes are anticipated or that unplanned unsystematic discoveries do not occur. Any effort to discern the unknown may lead to unexpected knowledge (accidental discovery). But without a systematic approach, one is left groping in random directions rather than exploring along a planned avenue.

2. Agricultural economics is viewed as a separate discipline by some—usually agricultural—economists. Those who view it as a separate discipline seem to base the distinction on the fact that agricultural economics did not evolve directly from economics. Instead it adopted its parent discipline after its initial inception. The author's inclination is to view agricultural economics as a subject-matter area of economics, but of a somewhat different status than other subject-matter areas because of its leading role in the development of empirical quantitative analysis in economics. Much of the development and testing of empirical methods, which is central to scientific validation of theory (Eichner, 1986), came first in agricultural economics.

3

Methodological Concepts and Perspectives

After a certain high level of technical skill is achieved, science and art tend to coalesce in esthetics, plasticity, and form. The greatest scientists are always artists as well.

Albert Einstein

Research methodology in economics is obviously not independent of the perspectives and boundaries of the discipline. It also conforms to certain long-established rules of the scientific and academic community. This chapter defines, discusses, and explains some general aspects of research methodology and its relationship to science, economics as a social science, theory, facts, objectivity, and such matters. The chapter begins by defining science, then progresses to considering knowledge, including classification of types, the role of personal objectivity, and the roles of facts, theory, and hypotheses.

SCIENCE

We may note that the term *science* has become associated with only the physical and biological sciences (chemistry, biology, geology, etc.) during the most recent century or so. McCloskey (1990b, p. 7) explains that "other languages use the science word to mean simply 'systematic inquiry.' . . . Only English, and only the English of the past century, has made physical and biological science into . . . 'the dominant sense in ordinary use.' " The treatment of science in this book is as a systematic inquiry rather than as a predesignated set of disciplinary areas.

Research is designed to generate new, reliable knowledge. The primary vehicle for achieving that reliable knowledge is science.

Science is the organized accumulation of systematic [reliable] knowledge for the purpose of intelligent explanation/prediction. (Williams, 1984)

36

The accumulated knowledge, the result of the processes of discovery and confirmation, is typically grouped into categories or branches of learning or instruction (e.g., chemistry, engineering, economics, physics, art, history), which we identify as disciplines or areas of study.

Several aspects of this definition are important. Note that the intended purpose of science is explanation/prediction. The slash is deliberately used because in the meaning of science, the two terms are intricately intertwined. By explaining what occurs and how or why it occurs, we develop the capacity for *conditional* prediction. That is, we are able to specify that "if X, Y, and Z occur, then W will follow." Scientific prediction is always conditional prediction. Unconditional prediction (e.g., foretelling the future) must have some means of knowing what conditions will be in effect (knowing that X, Y, and Z will, in fact, occur), and at least some of those elements are beyond the control or ability of the scientist to forecast with certainty. Thus, even when the conditions of a forecast are not made explicit, the predictions are still conditional.

While explanation is the primary goal, knowledge is the means of science. Science gives us the tools (accumulated knowledge) with which to do research, and research adds to the accumulated body of reliable knowledge. The two are mutually interdependent. Yet they are different, and the distinction is important. Science (or science*s*) as a body of accumulated reliable knowledge is an entity; but it is not a static or unchanging entity. Research is a process through which (1) the body of science is expanded; and (2) the existing body of science is tested for correctness or validity. The two are symbiotic—research draws on the body of accumulated knowledge as an integral part of its process, and the accumulated knowledge is expanded through the research process.

Research methodology is the methodology of science in the sense that it is the methodology of *expanding* science. Research is the means of accumulating, evaluating, and questioning bodies of knowledge we call science, as well as a process of using science in addressing issues, problems, and questions. Research and science are intricately inseparable. We distinguish between them because both are important separately, but neither can function without the other. As individuals, we cannot practice research without a grasp of existing disciplinary, subject-matter, and problem-solving knowledge (science), and we have no means to expand that knowledge except through a process of creating new reliable knowledge or understanding (research).

On the matter of science, we may as well dispel a popular misconception that "science" consists only of factual "truth," devoid of any human values and unaffected by private views. Simkin (1993, p. 12) states: "Metaphysics (i.e., philosophy or beliefs) and science have, of course, been closely related, both historically and logically." The Committee on Science, Engineering, and Public Policy (1995) states that science is inherently a social enterprise (p. 3), and that scientific knowledge emerges from a process shaped by human values, limitations, and social contexts (p. 2). The committee further notes that personal beliefs (philosophical, religious, cultural, political, and economic) can affect scientific judgment, but when judgment is recognized as a scientific tool, it is easy to see how science can be influenced by values (p. 6). Additionally, these social mechanisms and values both validate scientific knowledge and sustain methods and conventions for doing research (p. 4). This does not diminish the importance of science or its role in our pursuit of knowledge; it merely reminds us that scientific pursuits are carried out within social and individual systems of values, and by people who are fallible. We must recognize the role of individual, social, and scientific values in scientific pursuits and be diligent against blind acceptance (or rejection) of ideas, observations, or concepts. Put another way, science teaches both rational thought and freedom of thought (Feynman, 1999, p. 186).

Before progressing further into methodological matters, we will digress briefly into a discussion regarding economics as a science. While the views presented on the matter are not unique, they may serve to encapsulate the issues of a long-running and ongoing debate between economists and others—a debate that, although it will almost certainly continue, does not warrant substantial consideration.

ECONOMICS AS ART AND SCIENCE

To better understand this issue, some classifications of science are useful. Note that these are classifications, not definitions. We can break the sciences into the physical sciences (e.g., chemistry, biology, physics) and the social sciences (economics, sociology, anthropology, etc.), the distinction being that the physical sciences are concerned with things while the social sciences are concerned with people (Stewart, 1979, p. 5). The physical sciences are sometimes called the "hard" sciences. However, Johnson (1986b, p. 31) notes that they are really the "easy" sciences because they deal with inanimate things rather than human social behavior. Nevertheless, these groups, and the difference in focus between them, may be a key to understanding the debate.

The physical sciences were the first to adopt the designation of *science*, and they developed and evolved with a heavy orientation (belief?) in the philosophy of positivism (to be covered in chapter 4) and a strong empirical tradition—a reliance on and preoccupation with observations and data. Furthermore, in their developmental stages, several sciences evolved in a period and under a set of conditions in which the only feasible way to hold external forces constant was in a laboratory setting. Consequently, the laboratory, and the classical laboratory experiment, became their sole or primary "model" for being designated as "scientific."

The social sciences, on the other hand, study human behavior and phenomena that do not lend themselves to controlled or laboratory experiments. Controlled experiments in the social sciences are limited either because society does not find such experiments ethically acceptable or finds them prohibitively expensive. In place of laboratory-generated data, social scientists, and especially economists, have substituted and developed more sophisticated multivariate conceptual and statistical techniques to adjust for a myriad of forces that affect the major variables of interest. Further, economics is influenced by normativistic philosophy (also covered in chapter 4), which accepts social and individual values as being entities open to scientific inquiry. Positivists would accept the study of things as "real" because they have observable, physical dimensions, but values cannot be "real" because they have to do with states of mind. Those who exclude economics as a science because it fails to use a traditional laboratory or because of its range of subjects of study are defining science in terms of specific *methods* rather than in terms of the *methodology* (the general approach).

Another distinction between physical and social sciences lies in the relative emphasis placed on the balance between theory and data. While all sciences use both, a valid generalization is that physical sciences place more emphasis on data, while the social sciences place more on theory (these concepts will be defined more precisely and discussed further later in this chapter). Incidentally, these distinctions are not independent of the differences discussed above. The physical sciences generate data under controlled conditions as a means of testing their theories. These theories are, in simple terms, postulated relationships between inanimate forces, given specified conditions or other factors. As such, these sciences have a strong empirical tradition of verifying their theories through experimental design and data generation. The theories of the social sciences tend to address more complex phenomena of individual and group motivation and behavior and the effects of societal institutions on them. These phenomena are often not capable of being directly observed or quantified under artificially controlled conditions (Leontief, 1993 provides an insightful discussion of this issue). This does not, however, necessarily make them less real. On the other hand, some economists (e.g., Eichner, 1986) argue that economics has failed to confirm its theories empirically to the degree required by "science."

Having presented this view, we may also recognize that these generalizations weaken under extended examination. Consider, for example, that physics as a physical science relies more heavily on conceptualizations and somewhat less on controlled experiments than some other physical sciences. This is largely because many phenomena studied are complex, and controlled data are difficult. With regard to economics, Johnson (1986b, p. 82) notes that the criticism that economists are lax in using data to validate or test theory is directed more at the disciplinary work of economists than at the subject-matter and problem-solving work within the various fields of the discipline. McCloskey (1983, p. 491) states:

Economics is not a science in the way we came to understand that word in high school. But neither, really, are other sciences. . . . Economics is misunderstood and . . . disliked by both humanists and [physical] scientists. The humanists dislike it for its baggage of antihumanist methodology. The scientists dislike it because it does not in reality attain the rigor that its methodology *claims* to achieve.

McCloskey (1990a, p. 1127) also argues that economics is not sufficiently concerned with the empirical validation and testing (empiricism) to rank with the physical sciences, and that more focus on empirical methods in applied research would make it scientifically stronger.

Whether economics is an art or a science constitutes a stimulating philosophical topic for debate, but one's position depends on the initial premises one adopts. Economics is, like many disciplines, *both* art and science. It is a science in terms of its accumulated knowledge base, its descriptive knowledge, and its procedures. This includes economic research activities. It is both a science and an art in terms of applying that knowledge to current issues and problems. This includes both research and nonresearch (decision and action) types of activities. As economists, we know that the government setting a price floor on a good that is above the market equilibrium will cause a surplus of that good to accumulate, ceteris paribus. Yet the answer to the question, "What will happen if the price support for wheat in the United States is increased by 10%?" depends on how well we can anticipate what will happen to a host of other variables. These external variables may include levels of wheat production and income in other countries in the world, whether U.S. policy makers change other programs (e.g., food aid, export enhancement programs, acreage controls), and whether weather patterns in U.S. wheat-production areas are normal. However, in the application of science, the chemist also must anticipate effects of uncontrolled factors when predicting chemical reactions outside the laboratory. Thus, we all practice science to some degree within some boundaries, but include the practice of art, and become more multidisciplinary, when we attempt to make it applicable to problems and issues that are of concern in the larger "practical" world in which we participate. Research encompasses both art and science.

KNOWLEDGE

In considering reliability of knowledge, one of the problems is lack of agreement on what constitutes "evidence," but it is the reliance on evidence that constrains the influence of personal values in science (Committee on Science, Engineering, and Public Policy, 1995, p. 8). McCloskey (1983, p. 491) argues against constraining evidence to an artificially narrow range of criteria (e.g., the evidence must be "objective" or "experimental" or "observable"). In our examination of knowledge and its reliability, we will consider two classifications of knowledge, both of which are useful for purposes of examining research methodology. One classification system is by Larrabee (1964) and the other is by Johnson (1986b).

Positivistic vs. Normativistic Knowledge

The classification system by Johnson (1986b, pp. 16–20) specifies knowledge as positivistic knowledge and normativistic knowledge, with normativistic being subdivided into two types: prescriptive knowledge and knowledge of values. *Positivistic knowledge* is knowledge of conditions, situations, or things that are directly observable or measurable. *Normativistic knowledge* is knowledge about values. *Prescriptive knowledge* is knowledge of what ought/ought not be done to solve a specific problem. *Knowledge of values* is knowledge of the goodness and badness of conditions, situations, and things. The effects of the conditions, situations, and things are observable, but the inherent goodness or badness of the effects is a matter of perspective and interpretation. Knowledge of values includes knowing how people evaluate conditions, situations, or things in terms of good and bad, whether on some quantified scale or as binary entities. It can also be perceived as knowledge of what has value even though no one realizes it. When we determine, individually or collectively, that conditions, situations, or things are inherently good or bad, that knowledge is subjective. But knowledge of values is not inherently subjective; it may be the result of an evaluative, analytical, or measurement process and, as such, be as objective as positivistic knowledge.

Prescriptive knowledge is concerned with decisions about good and bad in a particular situation and inherently embodies judgment; it may be a logical extension of a combination of both value-free knowledge and knowledge of values. Prescriptive knowledge is necessary for individuals, groups, and society at large to make decisions and take actions. We all possess prescriptive knowledge, but it is conditioned by our own states of mind, personal and collective biases, experiences, perceptions of the world, moral/ethical conditioning, and other such factors.

We can make an additional differentiation within prescriptive knowledge. Prescriptive knowledge can be unconditional or conditional. *Unconditionally prescriptive knowledge* takes the form of "Prescription X is best (irrespective of the circumstances)." Persons claiming unconditionally prescriptive knowledge assume the attitude that their prescriptions are above question. Conditionally prescriptive knowledge is tempered by circumstances. It takes the form of "If the objective is Y, then the most effective approach is prescription X." *Conditionally prescriptive knowledge* recognizes that prescriptions are conditional on objectives and constraints.

Of these classes of knowledge, unconditionally prescriptive knowledge is handled cautiously in the research process. Recall that the goal of the research process is *reliable* knowledge, or that which is acceptable based on evidence. Reaching agreement on unconditionally prescriptive knowledge based on evidence is usually a severe task. This does not say that unconditionally prescriptive knowledge is "bad." All social entities must reach some agreements on unconditionally prescriptive knowledge (e.g., that first-degree murder requires a defined minimum penalty). However, those agreements are not reached through research. Research that deals with, for example, the effects of given actions, may contribute to those agreements, however. For purposes of our consideration of research methodology, references to prescriptions will be understood to be conditional unless otherwise noted.

It should be noted that there is no clear consensus among economists on the relative importance of conditionally prescriptive knowledge in the research process itself. For example, Edwards (1990, p. 5) maintains that prescription is *the* reason for doing research. Those who hold this view are focused primarily on policy or other decision or action applications of research. Those who confine themselves to disciplinary research may view prescriptive knowledge as less important. The differences in views about the role of prescription in research center on where to draw the line between research and decisions, particularly in cases when research is conducted with policy application in mind. This distinction is considered again later in this chapter and in chapter 6.

Private vs. Public Knowledge

Another way of classifying knowledge is as private or public knowledge (Larrabee, 1964).[1] *Public knowledge* is knowledge that can be demonstrated to others through logic or evidence. Public knowledge is, by definition, reliable knowledge. Reliable knowledge is public in nature; it is not depleted when shared (and often is expanded by sharing), and others cannot be excluded from its use (Stephan, 1996, p. 1200). *Private knowledge* is that which we accept or "know" for ourselves but that cannot be demonstrated to others. We all "know" (i.e., feel, believe, trust, etc.) things that we accept as reliable for ourselves but have no way to demonstrate to others. Perhaps the clearest example of private knowledge lies in the area of religious beliefs. Many of our individual or collective religious beliefs constitute private knowledge because there is no means to demonstrate the validity of the knowledge using logic or evidence. Acceptance of its validity rests on "faith."[2] It should be recognized, however, that knowledge in any field or discipline is private knowledge if its reliability cannot be accepted on the basis of evidence. Consider, for example, the knowledge that "capitalism is better than socialism." The reliability of this subjective proposition cannot be demonstrated because the criteria depend on one's view of what constitutes "better." On the other hand, the knowledge that "capitalism with competitive markets allocates goods and services more efficiently than does socialism" can be demonstrated by logic and evidence, given a specific definition of efficiency.[3] These distinctions will be further discussed in terms of subjective and objective statements and states of mind.

Private knowledge is used with caution in the realm of research. Note that private knowledge is not necessarily ranked as inferior, but we have no way to employ it to reach conclusions through research. Only public knowledge can be reliable in the scientific sense. However, private knowledge may lead to public knowledge when the circumstances allow it. When an individual intuitively understands something (possesses private knowledge) and sets about demonstrating the reliability of it, it may achieve the status of public knowledge. Much public knowledge likely started as private knowledge, and public knowledge (what is accepted on the basis of logic and evidence) is culturally and time dependent. Many of us have experienced this drive to demonstrate, show, or "prove" some aspect of our private knowledge to another person, the public, or the scientific community. We should recognize, however, that the desire to turn private knowledge into public knowledge has at least two major pitfalls: the potential incursion of subjectivity and the potential for the logical fallacy of special pleading. When we want something to be demonstrated because we have a vested interest in it, we are less likely to examine the evidence against it as carefully as the evidence for it and may, therefore, be more likely to present only the arguments that favor its acceptance.

WAYS WE OBTAIN KNOWLEDGE

We gain knowledge through six primary means: the senses, experience, intuition, revelation, measurement, and reasoning. While this list is not exhaustive, and they are not fully independent of one another, all avenues to knowledge can be placed under one of these categories, or some combination of them. The ensuing discussion attempts to briefly explain each.

   A.   *The senses.* Knowledge gained through the senses—sight, sound, touch, taste, and smell—may be regarded as the first step toward gaining private knowledge. It is also, as noted previously, the principal means of deriving value-free positivistic knowledge. As infants, we began accumulating knowledge through our senses, and we continue to rely on them for gaining knowledge about people and things throughout our lives. We might even classify this knowledge as "private facts." The things we taste, see, touch, and so on provide us with factual information. This knowledge may remain private rather than public because (1) our perceptions through our senses differ among individuals; and (2) the sensory information may not be capable of being demonstrated: there is nothing certain in our sense experiences because we interpret in light of what we have already learned (Simkin, 1993, p. 23). On the other hand, this knowledge forms part of the basis for each of the other avenues to knowledge except reasoning.

   B.   *Experience.* Experience constitutes the accumulated total of one's exposure to and interaction with people, places, things, circumstances, ideas, senses, and so on. Knowledge gained through experience may or may not be reliable, depending on its nature. Experience knowledge is essentially private until measures are taken to make it public. Experience, being what it is, is made of very divergent elements—a hodgepodge of information—and creating new reliable knowledge from it is typically a cumbersome process. Stated another way, it is a disorderly and unorganized means of learning (Williams, 1984). It is, however, often an essential component of making sense of knowledge that we possess or accumulate; we use our experience to evaluate new knowledge in its relation to prior knowledge.

C. *Intuition.* Intuition is the sensing or feeling of something being accurate or not. If it constitutes "knowing," it is knowledge in a vague form. As such, it is inherently private knowledge only. However, an intuitive understanding of a fact, relationship, or set of relationships may lead to an orderly exploration and logical development, which leads, in turn, to public knowledge. Back (no date) suggests that intuition may be a necessary condition for successful research, but it is not a sufficient condition for finding reliable knowledge. Intuition is often trustworthy as private knowledge and may lead to public knowledge, but in research and science it cannot be accepted on its own as reliable. As noted in chapter 2, Ladd (1987) is a strong advocate for the role of intuition in the creative research process, but he makes no claims that it can establish reliable knowledge on its own.

D. *Revelation.* Revelation, or the reception of knowledge from a vaguely defined or unknown source, is somewhat disquieting from a purely scientific perspective. It is not confined to divine revelation in a religious context. Most of us have experienced an idea "out of the blue": something that just occurred to or dawned on us. This is a form of revelation. It may spring from the unconscious mind and it probably relates to intuition, perhaps experience. Revelation as a source of knowledge is, however, not reliable. It may become reliable knowledge only if it meets the tests for reliability.

E. *Measurement.* Knowledge gained through measurement (quantification) is normally thought of as data. Its connection to the senses and experience is obvious. This is usually regarded as factual knowledge. It is also regarded as reliable most of the time, with the implicit understanding that there may be a sampling or measurement error associated with it. Some disciplines devote a large share of scientists' time and attention to acquiring factual knowledge through measurement. As suggested in earlier discussion, economics does not usually emphasize laboratory or field measurement of data in a physical sense, but statistical sampling and gathering data with surveys are important parts of economic research activity.

F. *Reasoning.* Reasoning may be viewed as the "final test" of knowledge or gaining knowledge. Knowledge obtained through reasoning is the only way to derive reliable knowledge of relationships and patterns through which we develop explanatory or predictive capability. Of the two avenues to reliable knowledge— measurement and reasoning—reasoning constitutes the only avenue to useful disciplinary, subject-matter, and problem-solving knowledge. Reasoning, being deductive, inductive, or both, is the way we establish the relationships, patterns, concepts, and basic theories through which we can make facts or data mean something that has relevance—to disciplinarians, managers, policy makers, consumers, voters, and so forth.

This does not say that reasoning alone can establish reliable knowledge, only that if knowledge cannot pass as logical it is not trustworthy. Ladd (1991, p. 4) points out that it is not possible "to prove anything about the real world by deduction because it is not possible to prove deductively that real types have the properties attributed to ideal types." Induction is fallible because hypotheses are derived from several assumptions, and observations that contradict a hypothesis indicate that at least one assumption is wrong but do not identify the wrong ones (p. 5). Consequently, both deduction and induction are fallible as tests of final reliability. Yet reliability cannot be established without one of them, often both.

RELIABILITY OF (PUBLIC) KNOWLEDGE

In our examination of knowledge and its scientific reliability, the question of criteria for reliability is critical to the notion of validity in a science or discipline (being both correct and meaningful). The criteria are confined to public knowledge. The criteria are also somewhat nebulous because they ultimately reduce to something on which reasonable people can agree. There are two general criteria for establishing reliability of knowledge: (1) it can be supported by evidence; and (2) the way the evidence is obtained or generated can be demonstrated or reproduced. This evidence may be quantitative (i.e., data) or it may be more complex logical constructions that include data but proceed on to relationships, generalizations, or deductions/inductions from the data. In the first instance, the observed phenomena may be measurable, or at least capable of being described or represented (e.g., photographed, drawn). We think of this as factual information or knowledge. But how do we establish the reliability of things that we cannot observe directly? We do it through reasoning, or logic.

Logic may be classified as being of two general types, deductive and inductive.[4] *Deductive logic* is the process of reasoning from general premises (e.g., assumptions) to specific results or conclusions. Economic theory rests largely on deductive logic. We establish a set or series of assumptions about conditions, motivations, and behavior, then logically work through to the expected outcome in terms of some set of variables or parameters that we wish to explain or predict. Theories of market behavior start with assumptions about market structure, people's motivations and objectives, and influence of external forces, then deduce a set of behaviors of market prices and quantities from those conditions. Mathematics may be the most clear-cut example of deductive logic. It begins with a set of definitions (of numbers, sets, etc.) and derives, through deduction, a large, complex system of theorems, laws, rules, and so on about behavior of numbers and systems of numbers.

Inductive logic is reasoning from the specific circumstances or outcomes to a conclusion about general circumstances or outcomes. It may also be represented as an empirical process of reaching a conclusion or arriving at new principles from known data and experience by observing objective realities (Ghebremedhin and Tweeten, 1994, p. 10). The most applicable explanation of induction for economics is statistical inference. By structuring a random sample of a larger population, collecting data about that sample, and analyzing the data using established criteria and procedures, we can infer characteristics and behavior about the entire population from the sample analyzed. Induction is used extensively in economics, especially in empirical research but in other ways as well. Additionally, the conclusions drawn from economic research—the generalizations about the larger world based on analysis or study of portions of it—are based on inductive reasoning.

LOGICAL FALLACIES

Whether the logic or reasoning is inductive or deductive, there are some logical fallacies, an awareness of which can make scientific inquiry (i.e., research) more efficient. We may note that to identify something as a fallacy does not say that it is automatically not so, but that it is not *necessarily* so. The description of what constitutes logical fallacies may vary with one's focus and intent. Different authors list them differently, some with Latin names, others with more descriptive terms (see, e.g, Stewart, 1979, pp. 20–29; Spencer, 1971, pp. 9–11).

The following list of logical fallacies is reasonably complete for our purposes.

A.   *Special pleading.* This logical fallacy occurs when one uses only the information that supports a predetermined position or conclusion, while ignoring other information. This logical fallacy is common in law and politics, even accepted as a desirable approach in trial law. Its pitfalls in the pursuit of reliable information or knowledge are obvious.

B.   *Affirming the consequent.* This logical fallacy occurs when conclusions are based on premises without examining the validity of the premises. An example is the unqualified conclusion that "free markets" always lead to an optimum social welfare. We know that optimum social welfare can occur under certain circumstances, but the existence of monopolistic power, market externalities, and other factors can negate that result.

C.   *Attacking the person.* This fallacy involves rejecting a position or conclusion because of attitudes about the person or group presenting the position or conclusion. It may take the form of "What individual X says is false because X is a Democrat (or Republican or black or Catholic, etc.)."

D.   *Appeal to the people.* This involves accepting or rejecting a position or conclusion because a large number of people accept or reject it. This fallacy has some elements in common with the fallacy of affirming the consequent.

E.   *Appeal to authority.* This fallacy occurs by automatically accepting a position, proposition, or conclusion because its source has specialized or in-depth knowledge of it. It may take the form of "Statement A is true because individual (or group) X says it is true." This fallacy also has some obvious common elements with affirming the consequent and can be viewed as a variation on the fallacy of appeal to the people.

F. *False cause.* This fallacy (also called the post hoc fallacy) is the danger of attributing the wrong cause to an effect. It reasons that "Because event B occurs after event A, A causes B." We are particularly susceptible to this logical fallacy from spurious correlation (when things occur together by sheer happenstance—there is no possible connection) or when another set of conditions causes both to occur. An illustration: California has the highest per capita consumption of wine in the United States and the highest incidence of blindness: therefore, wine causes blindness.

G. *Argument by analogy.* This fallacy reasons that "A is similar to B, so what is true for A is also true for B." A common application of this fallacious reasoning is that because certain government policies are effective, they also would be effective in China (or France or Zaire, etc.). Although by itself a logical fallacy, argument by analogy can also be a tool for discovery. That is, something that works in one situation may also work in another, but it cannot be *automatically* assumed to work. This fallacy also is related to affirming the consequent.

H. *Composition and division.* These are presented together because they are mirror images of one another. The fallacy of composition reasons that what is true for a part (an individual) is automatically true for the whole (the group). An economic example of this is "Since an individual consumer cannot affect the price of milk, all consumers collectively cannot affect the price of milk." The fallacy of division reasons that what is true for the whole is automatically true for each part. An example is that because a nation's income is greater when consumption increases, individuals' incomes are automatically greater when they consume more. These fallacies are well understood by economists because examples of them are explicit in economic theory.

TESTS FOR RELIABILITY

The above considerations apply to reasoning or logic alone. When we rely on other input such as data, observation, or empirical evidence (alone or in addition to logic) in the pursuit of reliable knowledge, avoiding logical fallacies is not sufficient. Furthermore, we usually use other input, particularly in applied research. Thus, for a more complete consideration of reliability, we need more comprehensive guidelines. When evaluating overall reliability, there are at least three tests that are recognized. These are the tests of logical coherence, correspondence, and clarity (sometimes called comprehensiveness) (Johnson, 1986b, pp. 43–49; Eichner, 1986, pp. 5, 6). We also can consider a fourth test—the pragmatic test of workability. Coherence is a logical test, constituting a necessary, but not sufficient, condition for reliability; the other tests are essentially empirical tests.

The *test of logical coherence* is examining an outcome or proposition to see if it is free of logical contradiction. This amounts to examining the outcome or proposition to see if it is free of logical errors, including the logical fallacies identified in the previous section. When empirical observation or data are involved, meeting the test of logical coherence does not ensure the reliability of the outcome, but its reliability fails if the test is not met. When applied to theory, it is a test of whether the conclusions follow logically from the assumptions.

The *test of correspondence* involves comparing an outcome or statement to what is already "known" to see if it is consistent with (corresponds to) prior knowledge. If a result is consistent with prior knowledge or results, it passes the test of correspondence. When applied to theory, it examines whether conclusions are supported by empirical evidence. Note that this does not say that we place unconditional trust in conventional wisdom. In fact, new findings that contradict what was previously "known" is a primary avenue to new knowledge. However, meeting the test of correspondence constitutes one piece of evidence to support a result, statement, or position.

The *test of clarity* examines the outcome or proposition for lack of ambiguity or vagueness. If the result or proposition has more than one meaning, it fails the test of clarity. Johnson (1986b) uses overidentified and underidentified reduced-form simultaneous model equations to illustrate a lack of clarity. Another illustration of lack of clarity is the development or application of a concept when the terms used in the concept are not defined, rendering their meanings ambiguous. When applied to theory, it is a test of whether the theory is consistent with all the facts pertaining to the theory (Eichner, 1986, p. 5). Unless a result or proposition meets the test of clarity, applying either of the other two tests may not be possible.

The pragmatist's *test of workability* (pragmatism will be addressed in more detail in chapter 4) requires that the result solve the problem or issue addressed. The outcome or proposition must achieve the desired result; that is, it must "work." The addition of this test as a test of reliability is relevant, of necessity, for economists engaged in problem-solving research, but not necessarily for those focused on disciplinary research. Thus, the view advocated here is that workability may be a useful test in given circumstances, but not as a test of reliability to be applied universally.

A simple example of these reliability tests may help to illustrate their uses.[5] Consider the situation of a physician giving a child a smallpox vaccination. Does the action meet the test of correspondence? For the physician, yes, because he or she knows that it has worked in all previous cases. For the child, probably no, because he or she lacks prior knowledge of the results. Does it meet the test of coherence? It does to the physician (who understands the concept) but not to the child (who does not understand how being hurt can help). The action meets the clarity test for the physician, who understands how the immunization occurs, but not for the child, who cannot grasp the process. The test of workability in this case is whether the child ever contracts smallpox.

In applying these tests, at least the first three must be met for us to view the concept, proposition, or outcome as reliable. Failure to pass any of the three (or four) tests discredits the knowledge. Furthermore, meeting these tests does not suggest that the reliable knowledge from the proposition is thereafter unquestionable; that is, passing all four tests is not "proof." It merely indicates that the knowledge is acceptable as working, reliable knowledge. All knowledge is continually under review, scrutiny, and question. Meeting these tests constitutes acceptable evidence that the information, thought, or knowledge is reliable.

While this classification of tests is useful for our thought process about reliability, a central point, or lesson, is that in the final analysis we must consider *all* of the relevant evidence when evaluating reliability of knowledge—disciplinary, subject-matter, and problem-solving. This point is widely recognized by scientists and philosophers (e.g., see Maki, 1996; McCloskey, 1983; Feynman, 1999; Committee on Science, Engineering, and Public Policy, 1995).

ROLE OF PERSONAL OBJECTIVITY

In contemplating matters and questions of science, knowledge and its reliability, and the role and function of research in the overall mixture of these entities, objectivity of the researcher/scientist is the lifeblood of reliable knowledge. Personal objectivity includes avoiding trying to impose our private values and beliefs on others. This does not mean that a completely objective state of mind must always be achieved, but that it be diligently sought. Complete objectivity is probably not a realistic capability of the human condition, but our best effort toward that goal is an aid in seeking reliable public knowledge.

The *subjective*, or subjectivity, is that which is conditioned by personal characteristics of mind; it is associated with beliefs, values, and opinion. Private knowledge is subjective if it is not demonstrable to others. Some things that are normative—what someone thinks should be—are also subjective. Personal judgments or prescriptions that are without logical or empirical support are both normative and subjective. However, all normative matters are not subjective. Knowledge of values is not subjective in a personal sense. The point of the distinction is that we cannot deal with personally subjective matters in science or research because there is no way to derive reliable public knowledge from personal subjective positions by themselves.

The *objective*, or objectivity, is that which is independent of characteristics of mind; it relies on what is demonstrable by observation, measurement, and logic. Things that are positive—identification or specification of what is—are also objective; they are free of personal values and beliefs. Issues, proposals, problems, and questions that are stated in personally objective rather than subjective terms are easier to deal with in the realm of public knowledge. For example, the question of whether the United States should lower import tariffs on a group of goods (a proposition stated subjectively) becomes arguable and is difficult to deal with in a research or scientific context (although often dealt with in a public policy context). Statements about the effects of lower import tariffs on some set of target variables such as domestic consumer and producer prices, national income, balance of trade, and payments are more easily debated as public knowledge. The second set of statements is in judgment-free language that constitutes a valid research question or objective.

Objectivity is widely recognized by researchers and philosophers of science as the cornerstone of science and reliable knowledge. Karl Popper, a well-known founder of the philosophy of pragmatism, emphasized that objectivity in observations, explanations, and criticisms is essential in the pursuit of science (Simkin, 1993). Feynman (1999) believes that the open, questioning, objective state of mind is the most important single element in the pursuit of reliable knowledge. He even argues that it is possible to follow the form and dictates of the scientific method and call it science when it is really "pseudoscience" without objectivity, and there is even a considerable amount of intellectual tyranny in the name of science (pp. 186–188). On the other hand, objective thought can become a habit of thought (p. 248).

What is the realistic capability of researchers and scientists to be personally objective? While that question has no answer in measurable terms, the quantification of objectivity is less important than an awareness of the concept as a goal. French (1971) argues that science has never claimed to be completely objective. We introduce personal subjectivity in our work even by deciding what issues, questions, and problems are more important as objects of study or research. If we remain on guard against the incursion of personal subjectivity into our research work, we produce more reliable research than if we are not diligent. More comprehensive perspectives on this are available in Johnson (1986b, p. 86) and Myrdal (1944, p. 55).

Randall (1974) offers the notion of scientific neutrality as a component of scientific objectivity, and this may be the real value of the goal of objectivity in research. Scientific neutrality embodies value neutrality, avoiding incursion of the scholar's values into his or her scholarly work. However, scientific neutrality cannot ensure neutrality of impact of the results of scholarly work. We must avoid "taking sides" on issues on which there are conflicting or competing interests, but research sometimes results in unequal distribution of benefits. Randall's central point seems to be that scholars and researchers cannot be partisan, though they should recognize that new knowledge has distributional impacts.

As a means of distinguishing between subjective and objective in the research process, the two following distinctions are useful: (1) a proposition or concept is objective if it passes the tests of coherence, correspondence, and clarity; and (2) a researcher or individual is objective if he or she is willing to subject statements to tests of coherence, correspondence, and clarity and abide by the results. Striving to be objective or neutral as researchers may be the most important single goal for us to work toward in the research process.

SCIENTIFIC PREDICTION

Recall that our consideration of concepts that are important for research methodology in economics started with a definition of science as the organized accumulation of systematic, reliable knowledge for the purpose of intelligent explanation or prediction. We then proceeded to consider several aspects of knowledge and establishment of its public reliability. We now will address some conceptual aspects of scientific prediction.

The layman's concept of prediction is typically one of forecasting future events without qualifying conditions on the forecasts. Don't we wish we had the power to perform that kind of feat! Perhaps some people have the ability (gift?) to foretell the future, but it is not within the realm of science or research to achieve such things. As stated earlier, scientific prediction is understood to be conditional because we have no way of knowing with certainty what all of the future conditions that affect an outcome will be. But if we understand how forces interact to produce a result, event, or outcome (i.e., if we can explain it), we have the basis for a conditional prediction. The conditional prediction does not provide as much as we may want, but it may provide the best that is realistically possible. It is also "intellectually honest" in the sense that it makes the conditions of the prediction explicit rather than hides them. For example, an unqualified prediction might be: "GNP next year will increase by 4%." The corresponding conditional prediction might be: "If all the economic sectors perform as we expect (as we assume they will perform), then GNP next year will increase by 4%." Economists are sometimes the focus of ridicule because of their propensity to make the conditions on which their predictions are made explicit (e.g., what we need is more one-armed economists).

The basis of scientific prediction is theory in conjunction with empirical knowledge, including experience. To examine and evaluate this singularly important proposition, we need to define some terms, then examine them and how they are related.

A fact is a verifiable observation.
A theory consists of logical relationships among facts.
A hypothesis is a testable proposition.

The reader is referred to Goode and Hatt (1952, especially pp. 7–17), for extended treatment of these concepts.

Facts constitute a portion of what was previously classed as reliable knowledge. They are observations that we establish through senses and/or measurement, and they can be demonstrated to and verified by other observers. Facts are positive—they establish what is, at least for a specified purpose; they are independent of personal value judgment. Facts, while they constitute reliable knowledge, are not necessarily permanent (Ladd, 1991, p. 8) and are not by themselves useful for prediction. This concept of a fact is different from the popular use of the term, which treats fact as "what is known to be true." This encompasses more than our definition of facts; it encompasses part of what we define as theory. The popular notion of fact is more a vague generalization (i.e., an impression) than a defined term with a specific meaning.

Note that the distinction between facts and values may depend on how a proposition is presented. For example, "Hunger is bad" is a value judgment, but "98% of the U.S. population believes that hunger is bad" may be a fact. Facts may be about knowledge of values. We can make four distinctions about facts and values.[6] There are (1) facts about facts; (2) facts about values; (3) values about facts; and (4) values about values. Items 1 and 2 are proper concerns of science, item 3 is suspect, and item 4 is outside the realm of science. The fact/value dichotomy is often oversimplified and misleading.

For our purposes, we can view facts as having two basic functions (Williams, 1984; Edwards, 1978). One is to initiate theory. Observations of outcomes (facts) in the world we observe gives rise to the need to explain how these facts occur— that is, the need to theorize. Second, facts assist in the reformulation, expansion, and clarification of existing theory. When we observe facts that are not fully consistent with existing theory, the discrepancies must be addressed.

Theory establishes relationships, using facts as building blocks, but is also based on deductive logic—reasoning from given (ideally, but not necessarily) factual premises to their necessary conclusions. The popular understanding of theory is that it is unfounded speculation or a premise without regard to reality or application (e.g., "That may be true in theory, but . . ."). WRONG! Theory is made of carefully thought-out relationships that hold when the specified conditions are met. The only test of a theory is its internal logic.[7] Theory is what gives us predictive/explanatory ability. It permits us to say: "If conditions A, B, C, and D exist, as we anticipate they will, then X and Y should occur." A theory is a model of typical initial conditions that bring relevant phenomena into typical relationships with one another (Simkin, 1993, p. 67). No given theory can possibly fit all circumstances. For a theory to be applicable to a given circumstance—for it to have explanatory capability for a given set of conditions—the premises (assumptions) on which the theory is predicated must be at least a reasonable approximation of the set of conditions to which the theory is being applied. When a specific theory does not fit a given current situation, that does not invalidate the theory; it indicates that it is not applicable for that purpose. Theories may be appropriate at a given time or under given circumstances but not others. Changing conditions can cause theories to become inappropriate or not applicable (Ladd, 1991, p. 8). Also, theory can suggest that a relationship is valid, but not whether it is empirically important (Edwards, 1978, p. 32). Theory cannot provide easy, "magic" answers or prescriptions for intricate and complex problems. Indeed, theories provide us with the framework with which to approach the analysis of these complex problems. Note that theory extends beyond simple causal laws; according to Simkin (1993, p. 65):

> Simple causal laws belong to an early stage of science. As the science develops into a theoretical system it becomes too complex for simple statements. The simple causal laws are limiting cases of scientific explanations that take the general form of probability statements.

Theory thus plays a central role in applied research by providing the conceptual base for specific applications. We may note that the influence also works in the reverse direction. Useful theory draws on reality and application. Theory may be developed in response to recognized real-world problems. If the theorist is not attuned to intended applications, how is he or she to know what to assume and include in the theory?

All theories are abstractions, simplifications, and/or generalizations. They necessarily must be abstractions or simplifications of reality because no reasoning process dealing with relationships can recognize or take account of every variable or condition that could ever affect an outcome. The theory would be so complex that it would be useless. It would apply to only one possible circumstance. Thus, all theory is generalization; it applies to a general class of circumstances. It is this characteristic of theory that makes it useful, because it can be adapted to different specific situations. Thus, theory gives us a framework within which to begin to reason through a given situation or problem to an expected outcome. This kind of application of theory gives rise to hypotheses about expected occurrences.

Heady (1949) identified five functions of economic theory in empirical research. These are to assist in (1) formulating the research problem; (2) formulating hypotheses; (3) designing empirical procedures; (4) assembling appropriate data; and (5) interpreting findings. The functions of theory are not limited to empirical research, however. The functions of theory in research have been addressed by Williams (1984), Edwards (1978, 1990), Ladd (1991), Castle (1989), and Breimyer (1991). Based on these works, the functions of theory in economic research may be listed as:

A.   *Orientation*. It provides a framework to examine or delineate a problem or question.

B.   *Classification*. It provides a precise means of communication. The set of definitions and specifications embodied in the development of theory provides terms and relationships that have specific, defined meanings that facilitate understanding of complex concepts and phenomena.

C.   *Conceptualization*. It gives the means to visualize how something works or to suggest cause and effect relationships that cannot be observed directly. Theory gives us the means to reason through alternate modes of behavior by abstracting to ideal types, giving them rules for behavior, and hypothetically playing them out.

D.   *Summarization*. This may be of two types:

1.   *Empirical generalizations* such as "Wood floats" or "Prices tend to be inflexible downward."
2.   *Generalized relationships*. An example is that the rate of growth in the money supply and the inflation rate are directly related.

E.   *Provision of precision* in thought processes. It helps to relate facts and concepts. Data may show correlations, but theory can help sort out cause and effect. Theory assists in thinking clearly and accurately but does not, in itself, provide criteria for solutions to problems or for policy. Problem solving requires other information, which makes it pragmatic and multidisciplinary/interdisciplinary.[8]

F.   *Prediction of facts or identification of hypotheses.*

G.   *Identification of gaps in our knowledge.* When the body of theory is inadequate to help explain phenomena, there is an inadequacy in our (theoretical) understanding.

The popular notion of theory is often close to our definition of hypothesis. A *hypothesis* is a result or outcome that is not yet evaluated or tested. It is a tentative assertion of a relationship between factors or events that is subject to verification or rejection. It is not necessary, sometimes not even useful, that a hypothesis be "true." A hypothesis must be capable of being accepted or rejected (rejected or not rejected in statistical hypothesis testing) on the basis of the evidence available.

Most researchers, economists included, automatically envision statistical hypothesis testing when the matter of hypotheses arises. In economics, hypothesis testing in econometric research, for example, typically involves testing hypotheses on two levels: (1) testing the statistical significance of individual parameters in the model; and (2) testing the statistical "fit" of the entire model, it being based on expected relationships specified in part or wholly by theoretical considerations. We might refer to statistical hypotheses, and hypothesis testing, as quantitative hypotheses. For enlightening discussions and perspectives on these hypotheses, see Tweeten (1983) and King (1979). Statistical or quantitative hypotheses must have three characteristics to be fully justified: (1) they must have a conceptual basis, that is, be built on theoretical reasoning; (2) they must be sufficiently specific to be rejected or not, based on the data; and (3) there must be data and techniques available to test them.

Another class of hypotheses, less explicitly recognized than quantitative hypotheses but at least as relevant, may be identified as qualitative or conceptual hypotheses (Tweeten, 1983). These hypotheses do not lend themselves to formal, quantitative evaluation. They are used instead as reasonable contentions, assertions, or premises. Although not testable in an empirical sense, they can be agreed upon or not by informed, objective evaluators. These hypotheses are grouped by Tweeten into three categories: maintained, diagnostic, and remedial hypotheses.

Maintained hypotheses are things assumed to be true for purposes of a study being conducted. The assumption that a research objective can be achieved is a maintained hypothesis, and the proposition that, for example, a specified market is structurally competitive is a maintained hypothesis. Acceptance of a maintained hypothesis in the form of general agreement is necessary for the study to proceed and for the results to be accepted.

Diagnostic hypotheses are propositions about the causes of a problem. For example, reasoning that "cheap food" policies of developing countries contributes to low agricultural productivity constitutes a diagnostic hypothesis; alternatively, reasoning that wage controls will lead to labor market inefficiencies in a particular instance is a diagnostic hypothesis. As with maintained hypotheses, acceptance of diagnostic hypotheses as a consensus of analytical reasoning is necessary to establish the premises of a study or the acceptability of results and interpretations. Yet neither is necessarily empirically testable.

Remedial hypotheses are proposed solutions to problems. As Tweeten notes, they are often related to diagnostic hypotheses. They offer prescriptions for potential solutions to problems for which diagnostic hypotheses identify the causes. An example of a remedial hypothesis to the above diagnostic hypothesis example is the prescription: "Removing price controls in country Z would increase food production within the country to a domestic subsistence level." Remedial hypotheses are often conditional ("The goal of X can be achieved by action Y"), and they frequently appear in the implications or conclusions of research. These hypotheses, as with the other two, are typically not empirically testable but are accepted or rejected by weight of evidence by reason.

SUMMARY

This chapter provides definitions and concepts of science, knowledge, theory, and personal objectivity and examines their relationships to research methodology in economics. Science—the organized accumulation of systematic, reliable knowledge—provides a knowledge base for research. Research, in turn, adds to the body of science and is the vehicle for validating scientific knowledge. Economics is presented, along with other sciences, as being scientific in its discipline knowledge and research methodology. Extending the science to multifaceted contemporary societal problems and issues often requires multidisciplinary/interdisciplinary research, judgments and decisions.

Knowledge is classified in two ways: (1) positivistic and normativistic (prescriptive or knowledge of values); and (2) private and public. Both are useful in evaluating and understanding knowledge generated through research, although only unconditionally prescriptive and private (personally subjective) kinds of knowledge are not valid within the context of research activities. The basic avenues for obtaining knowledge—the senses, experience, intuition, revelation, measurement, and reasoning—are examined in relation to *reliable* knowledge. Tests for reliability of knowledge are the tests of correspondence, coherence, and clarity, and some wish to add the pragmatic test of workability. In a more general, strictly logical vein, eight logical fallacies are also presented.

Personal objectivity as a state of mind of researchers is a most important goal in economic research. The role of objectivity is to avoid bias in the production of knowledge, not to deny the place of values, beliefs, and opinions in one's activities.

Facts, theories, and hypotheses are defined and the roles of each in science and scientific prediction are addressed. The relationships of facts, theories, and hypotheses to each other is covered, their functions identified, and popular misconceptions about them are discussed. Various classes of hypotheses—quantitative and qualitative—are also introduced.

RECOMMENDED READING

Johnson. *Research Methodology for Economists*, pp. 16–20, 43–49.
Larrabee. *Reliable Knowledge*, pp. 1–24, 104–106.
Randall. "Information, Power, and Academic Responsibility."

SUGGESTED EXERCISES

1. Describe a specific instance of an application of economics in which it represents "science," then describe a specific instance of an application of economics as "art." Do the same thing for a physical science (biology, chemistry physics, geology, etc.).

2. Does one's political philosophy classify as public or private knowledge? Explain.

3. Describe an example of each of the eight logical fallacies based on your own experience.

4. Select a journal article from any recent issue of an economics journal. Identify the first five subjective statements in the article and (a) explain why/how each statement is subjective; and (b) change it to an objective statement.

5. Describe an example, from your own experience, of a situation in which different people have (or would have) taken the same premise or proposition and interpreted it alternatively as comprising a fact, a theory, and a hypothesis.

6. Provide a written description of each of the three types of qualitative hypotheses identified in this chapter.

NOTES

1. This notion may be related to Johnson's (1986b) thoughts on interpersonal validity (pp. 241, 242), but Larrabee's discussion is much more extensive.

2. Johnson (1986b, pp. 228–229) argues that all descriptive knowledge rests on "faith" because it cannot be proven.

3. Unless efficiency is clearly defined as to its criteria, the statement remains a normatively stated proposition because "efficiency" presumes certain objectives. For elaboration of this point, see Ladd (1983, especially pp. 1–3).

4. Some maintain that all logic is necessarily deductive, and this precept is accurate by most historical standards. However, induction such as that embodied in statistical hypothesis testing is inherently as logical as deductive reasoning. It may be noted that the theoretical basis of statistical tests relies on deduction (reasoning from premises or assumptions to conclusions). Thus, on careful examination it may not be clear whether the basis of the reasoning is inductive or deductive. Nevertheless, inductive logic is used here to be synonymous with inductive reasoning.

5.   The author is indebted to Josef Broder for this example.

6.   This perspective was provided by Josef Broder.

7.   Some maintain that the conclusive test of a theory is its ability to predict. However, this confuses the logic of a theory with applicability to a particular situation. That is, when a theory does not explain a given situation it does not apply to that situation, but the theory may explain other situations. The theory of perfectly competitive market behavior is a useful example. Just because that theory does not explain pricing behavior in the U.S. automobile industry does not invalidate the theory of perfectly competitive markets. The premises of the theory (the conditions for perfect competition) are not consistent with the conditions in the U.S. auto industry. Thus, that theory is not applicable to that particular market situation. But the assumptions of the theory are close enough to the conditions existing in the wheat-producing sector of the U.S. economy that competitive market theory is useful in explaining behavior in that industry. Eichner (1986, p. 5) notes that economists reject the view that theories must be empirically confirmed (a view held strongly by positivists). Within economics, empirical confirmation is valued more highly by applied than by disciplinary economists. Simkin (1993, p. 16) argues that demarcation of theories according to testability is not the same thing as demarcation according to intellectual worth.

8.   The distinction between multidisciplinary and interdisciplinary research is as follows. Multidisciplinary research uses concepts and/or tools from two or more disciplines; the different disciplines contribute to the research process, usually with researchers from the separate disciplines working together in tandem. Problem-solving research abounds with examples of multidisciplinary research. Interdisciplinary research goes further to actually integrate concepts from different disciplines in the research process. Subject-matter research offers many interdisciplinary examples. Consider that environmental or natural-resource economics often integrates biological concepts with economic concepts in research in that subject-matter field.

4

Philosophical Foundations

Economic thinking and research methodology in economics have been and will continue to be influenced by different philosophical positions held by individuals and by groups of individuals with similar views and beliefs. Each philosophical position holds views about and offers prescriptions for what is "proper" in terms of activities, attitudes, and approaches used in economic research. There are many philosophical value positions that might be identified, but we will focus on three of them—positivism, normativism, and pragmatism—that have had a major role in shaping economic thinking and ideas of conducting research in economics. Note that the labels for these philosophies may vary, and suffice it to say that there is considerable disagreement among individual economists, and groups of economists, on these philosophical positions and the relative importance of each in its effect on economic thinking. In this chapter an attempt is made to synthesize or generalize about these philosophical approaches. The intent is not to choose among them or to advocate one over another, but rather to introduce and illustrate how these philosophies have contributed to methodological approaches in economic research.

One of the reasons why we need to understand something about philosophical positions is to be able to deal with the beliefs of others as well as our own attitudes and beliefs. The following, from a personal communication with George W. Ladd, is offered to provide perspective on this point:

> There are some aspects of economists' behavior that I could not understand until I studied philosophy of science. I will illustrate with a personal experience with two economists—call them A and B—whom I have always respected and still do. Initially A did not respect B, and I could not understand why, but B respected A. Now each respects the other. B does problem solving or subject matter research (Johnson's terminology); most of it is normative or conditionally normative; much is multidisciplinary. A has always done strictly disciplinary research that is positivistic, and he adheres to Friedman's philosophy. When he was younger, A thought that economists should only do positivistic disciplinary research, and therefore believed that B was not being an economist. B, however, has always recognized that the profession can profit from different philosophies and different kinds of research. As A has matured, he has come to share B's broader view of the profession.

One's philosophical beliefs affect one's choice of "legitimate questions" (i.e., of one's selection of issues that are proper for economists to study) as well as one's choice of research methods.

The importance of methodology, including its dependence on philosophical beliefs, was also stated succinctly by Randall (1993, p. 58) as follows:

> Because researchers are the front-line methodologists for their own research programs, it follows that the serious study of methodology and rhetoric is desirable for researchers. Such study is unlikely to yield hard-and-fast rules for research, but is likely to yield a rich lode of insights. . . . Furthermore, it tends to make researchers more introspective about their own work, which is all to the good.

The nuances of philosophical arguments and views are complex and there are rarely, if ever, simple outcomes to the issues. Different philosophies as they relate to science and research represent different views of what constitutes reliable knowledge, or what is proper activity or procedure for establishing reliability. Each of the general philosophical positions addressed here offers views that are well founded, yet each has premises that can be questioned. This chapter does not resolve any of the conflicts or inconsistencies. But as students of research methodology in economics, we need to understand at least the basic elements of the general philosophical positions and disagreements as we attempt to conduct our research activities, and evolve our own philosophical positions. Remember that the boundaries of research activity are not set by any single philosophical position.

The three philosophies most connected to research methodology are positivism, normativism, and pragmatism. In the material that follows, the ideas and concepts underlying these philosophies are introduced. For a more thorough treatment of them, readers are referred to Glenn Johnson's book (1986b, chs. 4–9) and should consult the references he provides. That material represents an extensive and insightful explanation of the three philosophies, and the summary below draws heavily on Johnson's work. The serious student of the philosophical underpinnings of economic methodology, including research methodology, must read far beyond the treatment of the philosophies presented here.

The philosophies presented are philosophies in the context of confirmation, not in the context of discovery.[1] Confirmation is concerned with determining the reliability of knowledge, while discovery is concerned with the finding of new knowledge in the first place. Confirming the reliability of knowledge is the central concern of the methodology of science, and the philosophies of positivism, normativism, and pragmatism hold different positions or views on how that may be done. This does not say that science is not concerned with discovery, but rather that the philosophical positions do not directly address the process of discovery.

The methodology of research is concerned with confirmation, but it also is concerned with discovery, although the prescriptions for achieving it are nonspecific. We understand that discovery embodies creative thinking, for which the techniques are less structured, while we have more highly developed, structured techniques and procedures for confirmation. The act of discovery may require relatively more creativity and intuition than does the act of confirmation. While this book attempts to encompass both confirmation and discovery, the caveat offered here is that the three philosophies discussed in this chapter are concerned primarily with confirmation, not discovery.

As a prelude to the discussions of positivism, normativism, and pragmatism as central philosophies, a cautionary note may be useful. Do not assume, as some may suggest, that science and research are inherently associated with a particular philosophy. There are still some in the sciences, usually in the physical sciences, who believe that positivism is the only valid philosophy underlying the "scientific method." However, McCloskey (1983, p. 482) makes the case that this concept of the scientific method does not describe the sciences as it was once thought, and that purely positivistic models are not necessarily good models for economics anyway. The position taken here is that all sciences and the collective body of research embody elements of each of these philosophies.

POSITIVISM

Positivism as a philosophy adheres to the view that only "factual" knowledge gained through observation (the senses), including measurement, is trustworthy. "Pure" positivism even discounts reasoning and theory as valid for arriving at reliable knowledge. However, the more modern view of positivism embraces the logical extension of the facts. This is called logical positivism, and it became the predominant positivistic philosophy early in the twentieth century; few in the scientific community subscribe to pure positivism today. Logical positivism is the point of reference henceforth in this discussion.

Positivism originated in the physical sciences and some in those disciplines still view it as *the* philosophy of science even though it is declining in importance in the scientific community relative to the other two philosophies. Logical positivism has influenced economics, particularly in the twentieth century, although it has been waning relative to pragmatism, and has never been the dominant philosophy in economics. Logical positivism was to become influential in economics by the 1950s with proponents such as Wassily Leontief, Milton Friedman, and Harry Johnson (see Friedman, 1953; Johnson, 1975; Leontief, 1993). The philosophy had been building among economists since Alfred Marshall's time. Johnson (1986b) points out that John Neville Keynes, John Maynard Keynes's father, had been an early proponent of logical positivism in economics in the 1890s. The logical positivism of economics may be viewed to include the study of society's values as well as descriptive, value-free knowledge, but some in the physical sciences would not regard any knowledge about values as positivistic.

The philosophy of positivism disavows the reliability or scientific validity of both prescriptive knowledge and descriptive knowledge of "real" values. Logical positivists will accept the validity of knowledge about values people hold regarding things or circumstances (e.g., that people believe that cancer is bad) but will not accept the validity of descriptive knowledge about values as real characteristics (e.g., that cancer is bad irrespective of beliefs). That is, positivists accept the "measurement" of people's or society's beliefs about what is good or bad, but not the knowledge that anything is inherently or descriptively good or bad. Concepts of good/bad or right/wrong are viewed by positivists as constructs of the mind, not as describable characteristics of reality. The view of the positivist is that only things that can be directly observed or measured are valid for scientific attention, and things that are not observable or measurable are not meaningful to the positivist; some philosophers of science use the term *standard empiricism* to describe positivism (Maxwell, 1992). In positivism, the emphasis on experience is so encompassing that there is no recognition of the significance of a practical or "real-world" problem as a reason to observe and assemble facts and postulate hypotheses (Back, no date). Lewin (1996, pp. 1305, 1306) states:

> According to logical positivism, science does not seek to help us to *understand* the nature of reality; that is the domain of metaphysics. *Understanding* is . . . a meaningless pursuit. The world simply consists of observable empirical regularities, and science should restrict itself to *describing* these, in the form of objective, falsifiable propositions. Every scientific concept and theory should have a clear empirical meaning; no ambiguity can be tolerated. . . . Therefore, if economics is to be a science, then we must transform the tautological propositions of pure theory into falsifiable scientific propositions, by assigning precise empirical meanings to all theoretical concepts.

However, positivism, like all philosophies, is based on empirically untested presuppositions. Positivists reject untested presuppositions that good or bad is empirically knowable but accept the untested presupposition that experience is empirically knowable. This position may "grow out of the failure to distinguish between the good and bad and the right and wrong (which are the logical consequences of decision rules and therefore not part of the natural world to be observed)" (Johnson, 1986b, p. 43). Positivistic knowledge is also culturally dependent because the experience through which we interpret observations is affected by the surrounding culture.

Castle (1989, p. 3) notes that logical positivism holds that theoretical concepts are valid only if the theory or its propositions can be quantified. This position is also too extreme a position for many economists, although we would prefer to include quantification whenever possible. Quantification is not always possible, however. Another point worth noting is that logical positivists tend to see both facts and theory as sources of hypotheses. Economists embrace this component of positivism, seeing both theory and facts as contributing to hypotheses, often by interacting; this is discussed further in chapter 8.

Economists cannot fully embrace the philosophy of positivism. This is largely because many things that are not concrete are nonetheless "real." For example, no one has ever directly observed a demand relationship, but its conceptualization and representation are real and its characteristics can be estimated; it is logically positivistic. One does not necessarily have to be able to touch something for it to be real. A commodity demand relationship is, to us, as real as the physical existence of a chemical element. Because a value must be "seen" through a logical process of conceptualization does not detract from its existence. Some things can be seen only with the aid of an electron microscope; some things can be seen only with the aid of reasoning. Thus, the senses are not the only means through which experience leads to information or knowledge.

While economists do not endorse the philosophy of logical positivism to the exclusion of other philosophies, it has had a profound effect on economic thinking and economic research. One way it has influenced economics is to place more emphasis on measurement and quantification when possible. Empirical analysis and estimation get at least part of their impetus from the thrust to be more positivistic. New methods in statistics and econometrics are influenced by positivistic thinking. However, positivism and empiricism are not synonymous. Empiricism includes measurement of things and phenomena but extends into quantification and estimation of relationships. In fact, Caldwell (1982, pp. 19–35) characterizes logical positivism as evolving toward or maturing into something he characterizes as logical empiricism, which is a sort of merging of logical positivism with pragmatism.

Another major way in which positivism has influenced economics is to focus attention on knowledge about values as being positivistic knowledge, while prescriptive knowledge is substantially different. For example, behavioral attributes of people, institutions, or groups are positivistic when they are quantifiable and/or demonstrable. Identifying, quantifying, or demonstrating what people value, or how much they value things, is little different from positivistic knowledge about physical things that can be observed or measured.

A third way in which positivism as a philosophy has had an effect on economic thinking is to highlight the importance of objectivity in the practice of economics and economic research. That is, positivistic philosophy, in emphasizing the importance of providing evidence to support a descriptive conclusion, argues the nebulousness of personal judgments and perceptions. A paradox of this stance, however, is that judgments are inherent, not just with respect to value, but with respect to "facts" or "reality." Rudner (1953, p. 2) explains that scientists inherently make value judgments about things like confidence intervals on descriptive parameters. When we set confidence limits for statistical tests, we make a decision about what level of evidence is *sufficiently* strong to warrant not rejecting a hypothesis. He proceeds to argue that "objectivity for science lies at least in becoming precise about what value judgements are being and might have been made in a given inquiry" (p. 6). To cast it slightly differently, perhaps the useful admonishment of positivistic philosophy, and our consideration of it, is to remind us to be on guard to distinguish between private and public knowledge (i.e., between what we accept as valid in our individual private lives and what we can convincingly demonstrate to others using experience and logic).

The tests of logical positivism are the tests of correspondence, coherence, and clarity discussed in chapter 3. These tests fail to recognize, as we shall discuss later, any consideration of usefulness or applicability. This makes the philosophy of logical positivism incomplete in a sense but does not damage the philosophy as a part of a more general approach to research in economics.

The behavior of economic agents and entities and their valuation of things is a legitimate part of positivism, although nonbehavioral aspects are not because positivists believe we cannot measure values as such. Positivistic philosophy affects disciplinary and subject-matter research but does not accept the prescriptive objectives in problem-solving research in economics. Its role in disciplinary research is seen in its effects on empirical testing in the development, refinement, and validation of theory, in disciplinary knowledge, and in the research on new procedures for studying relationships. In subject-matter research, its effects are most obvious in the estimation of parameters and relationships that provide the basis for policy and management decisions, even though positivists would disavow the prescriptions required of the decision process.

Summarizing positivistic philosophy, it has had important and lasting effects on the way we think about science, research, and reliability of knowledge, but is no longer the dominant philosophy of science. It is too limiting to be dominant, especially in economics but also in many other disciplines. The criticism of positivism as a philosophical position may have been most succinctly stated by a physicist (Heisenberg, 1991, p. 826):

> To positivists, the world must be divided into that which we can say clearly and the rest, which we must pass over in silence. But it is a pointless philosophy, because we would be left with trivial tautologies.

But irrespective of these criticisms, Johnson (1997, p. 262) makes this case:

> Logical positivism has and still does provide philosophic orientation for a considerable part of mainstream determinate disciplinary economics and it continues to guide the work of most physical and biological scientists, even if it is being subjected to increasing criticism from leading intellects.

NORMATIVISM

Normativism as a collection of philosophies (Johnson, 1986b, p. 34) related to science takes the position that knowledge of the goodness and badness of conditions, situations, things, and actions is valid and even necessary in order to produce prescriptive knowledge. Some forms of normativism, such as utilitarianism, hold that good and bad are experiencable (e.g., cancer *is* bad). Some hold that descriptive knowledge of intrinsic value of good and bad is possible and that this knowledge can be used to draw prescriptive conclusions about what is right and wrong or of what should or should not be done. To normativists, conditions and situations can be good or bad, but they cannot be right or wrong. Actions can be both good or bad and right or wrong. The view presented here is that normativistic philosophical beliefs are more prominent in the broad range of problem-solving and subject-matter economic research than in the more restrictive activity of disciplinary economic research within economics. Normativistic philosophy in economic research emphasizes matters on which people place value—things such as efficiency, welfare, income, standard of living, and quality of life. To the normativist, as the research relates to activities like business and policy decisions and recommendations, goodness and badness considerations become more central to the prescriptive process. Normativists accept the presupposition that intrinsic values (good and bad) are knowable, if only from a given perspective. They accept, however, that good and bad are as they are because people believe them to be.

Normativism as it applies to economic research is not directly concerned with moral questions of or moral prescriptions on right and wrong, which are fundamentally different from questions or knowledge of what is good and bad (Lewis, 1955). However, we often take as given the societal rules that govern societies and the markets that function within them, which are essentially moral/ethical rules (Johnson, 1986a, pp. 73, 74). Economic choices may, perhaps often, entail selecting between several "bad" alternatives; the choice of a "right" alternative under the circumstances may not involve a "good" choice or outcome. Whether a condition, situation, or action is right or wrong in a moral sense may not be a concern within economics. Thus, when we talk of normativistic philosophy as it relates to economics and to economic research, we do not encompass *all* normativistic philosophy. The purpose is to avoid the incursion of personal private values and maintain the focus on public rather than private knowledge of both intrinsic and extrinsic values. This, consequently, places emphasis and importance on the objectivity of researchers.

Objective normativism[2] (Johnson, 1986b, pp. 57–62) refers to a position that the desirability (goodness) of a result or outcome (or lack of it) from the point of view of individuals or groups can be known (e.g., based on observation or experience and logic), whatever the beliefs are and for whatever reason they may exist. Situations, actions, and so on can then be analyzed in terms of their goodness/badness given those views (values). Objective normativism as a philosophy is not very different from the philosophical position of logical positivism. Johnson (1986b, pp. 59–60) states:

> With respect to value-free positive knowledge, it is important to recall that methods used in acquiring such knowledge involve the subjectivity of sense impressions, the necessity of interpreting sense impressions, leaps of faith

from sense impressions to reality, insight, intuition, and the need for interpersonal acceptance of perceptions of reality. This leads to the position that objective descriptive knowledge about the value of conditions, situations, and things is fundamentally similar to knowledge on the value-free positivistic side, and that both value-free positivistic and value propositions are judgmental in nature. We also note that on both sides we have to put up with both probability distributions and the possibility of making mistakes in interpreting our sense impressions and in employing our logic. Further, it should be kept strictly in mind that we are only considering knowledge of values here and are excluding prescriptive knowledge of rightness or wrongness.

Objective normativism accepts that objective value knowledge is sometimes essential for statements or prescriptions about what should be done to accomplish specified goals or objectives. For example, "The people of the world are better off with free trade than with restricted trade" is an objective normative statement. The statement, "If the goal is to increase farmers' incomes with minimum impact on the functioning of commodity markets, a program of direct income supports rather than one of commodity price supports constitutes the better policy" is a conditionally normative statement that also qualifies as objectively normative (assuming it is based on sound analysis) because the assumption (condition) is explicit. If the statement becomes, "If the goal is to increase farmers' incomes with minimum impact on commodity markets, the government should adopt a program of direct income supports" it is objectively normative *and* prescriptive.

Normativistic philosophy is an inherent part of economics and economic research, and there is no reason to consider it inferior or to apologize for it—as long as it is kept objective as contrasted with subjective (which is necessary for both value-free and value knowledge). In fact, it is the element of objective normativism that makes the discipline, and our research processes and results, relevant to most of our clientele. These include policy makers of various types, with different concerns and problems, and firm managers and industry leaders. Our clientele also include the general public with an array of concerns for issues ranging from environment and living standards to international trade and national defense. In fact, any discipline that deals with public policy *must* use normative value judgments, and economists/economics must, by definition, be concerned with values, both monetary and nonmonetary. We believe that economics and economic research have something to offer everyone on matters that concern them. That capability would be limited without elements of normativism embodied in the attention of economists. What we must remember is that we are not in a position to impose individually subjective perceptions of value on others.

Objective normativism, as applied in economics, contributes to all three types of knowledge. Its connection with knowledge of values and prescriptive knowledge is direct and obvious. The connection with positivistic knowledge is indirect—through knowledge of values—and is dependent on one's acceptance of the argument that objective knowledge about values is no different from value-free positivistic knowledge. This philosophy influences disciplinary research through its focus on knowledge about values and behavior. It is exhibited in subject-matter research through both knowledge of values and prescriptive knowledge. Problem-solving research depends more heavily on the prescriptive knowledge emphasis of objective normativism.

PRAGMATISM

Pragmatism as a philosophy holds that what is important with respect to descriptive knowledge is how well it works for solving a problem at hand. Pragmatists are interested primarily, if not completely, in prescriptive knowledge. Attention to theoretical logic for its own sake and distinctions between abstractions such as positivistic knowledge and knowledge about values are relatively unimportant for pragmatists. They view positivistic value-free knowledge and normativistic value knowledge as being interdependent and place little confidence in the distinctions between them or even believe such distinctions are impossible. Pragmatists also may see means of obtaining knowledge and the ends producing the knowledge as interdependent. For example, a prescriptive result of a research process may produce an understanding of consequences that leads to a shift in the desired result. Pragmatists evaluate concepts for their usefulness in solving contemporary problems rather than for their own sake. That is, pragmatists are interested in *applying* concepts to solve problems. As part of this focus on solving problems, pragmatists regard value-free knowledge and knowledge of value as interdependent and inseparable. Workability (appropriateness for the problem at hand) is a central pragmatic criterion for judging empirical propositions (Johnson, 1986b, p. 37).

An example may be useful to help understand the pragmatic interdependence of value-free and knowledge of values. Consider the hedonic price analysis studies done on automobiles, with the objective to determine the market values of attributes or amenities of automobiles rather than the values of automobiles as a composite of attributes (see Griliches, 1961, for example). These studies used data on the attributes of automobiles (e.g., colors, engine types and sizes, existence of air conditioning, and many other attributes), which consisted of value-free knowledge, and data on prices of the automobiles (knowledge of values) to impute the values of the attributes embodied in the autos. Distinguishing between the two types of knowledge was immaterial, and the interdependence was necessary to produce results. The results provided prescription for adjusting price indices for quality change that occurs through time.

Pragmatism as a philosophy affecting economics became a prominent force with the institutionalists in the 1920s. While some assert that the institutionalist school of thought has had little important influence on formal, theoretical economic thought, the matter is certainly debatable. Pragmatism has had a substantial, and perhaps increasing, influence on research methodology in economics. Institutionalists emphasized the importance of society's institutions in economic outcomes, and economists today are more attuned to those roles, as evidenced by the incidence of Nobel Laureates for Economics going to students of public choice and institutional economics. The lasting influences of the pragmatic institutionalist perspective has been to focus economics more on solving problems and to make the economics of institutional change part of the mainstream of economics.

From the pragmatist's perspective, the test of workability is the primary test for reliability of prescriptive knowledge. If a concept works prescriptively (solves the problem) then it is acceptable. However, tests of clarity, coherence, and correspondence also have a role in pragmatism even though the test *may* not be applied before the prescription is implemented. Clarity is achieved if prescriptive solutions to problems are not ambiguous or vague. If the prescription is valid (it works), coherence (absence of contradiction) is achieved (Johnson, 1986b). If the outcome is consistent with what we already know, it meets the test of correspondence.

Pragmatic philosophy has affected economic thinking and research in profound ways, especially in subject-matter and problem-solving research, and especially in the United States during the twentieth century. Possibly the most significant influence has been to place the focus on nondisciplinary problems/issues as reasons to initiate research inquiry. Northrop (1959) attributes this pragmatic focus for research primarily to John Dewey, an early-twentieth-century philosopher. Within economics, pragmatism is a more prominent philosophy with some individuals, groups, and institutions than with others. Generally, pragmatic philosophy plays a larger role in problem-solving research than in subject-matter research, and has the least influence in disciplinary research. Even so, pragmatism plays a major role in most aspects of economic research because much of our focus is on either management or policy issues.

All "decision sciences" embrace pragmatism in some form or to some degree. U.S. educational and political systems are dominated by a philosophy of pragmatism. We are conditioned to be concerned with problem solving and workability (achieving results) in our educational system and through political activities and processes. Most Americans take pride in pronouncing themselves pragmatists; we economists from this culture cannot avoid being influenced by that conditioning. All our research that has potential policy or management applications draws on a pragmatic interest in applicable results/findings. The institutional school of economic thought, the only school identified as distinctly American, is based largely on pragmatism. However, while pragmatism as a philosophy has had a major influence on economics, its effect has been partial and selective, as has been the case with the other two major philosophies. As with the other philosophy positions, pragmatists make their own presuppositions. For example, they accept the presupposition that the difference between knowledge of values (both intrinsic and extrinsic) and value-free knowledge does not matter. They also accept the presupposition that *why* prescriptive knowledge is valid is secondary to concerns that it is valid.

HOW THE PHILOSOPHIES BLEND

Most disciplines and fields of study have adopted and adapted the basic philosophies of logical positivism, objective normativism, and pragmatism in some combination. There are also differences in individuals' philosophies within a given discipline or field of study. For example, mathematicians who are interested in the problem-solving applications of mathematics philosophically embrace pragmatism and/or objective normativism to some extent. But many mathematicians have a low tolerance for pragmatism, certainly compared to most economists. While recognizing these inherent differences, it is nevertheless useful to generalize to some extent about how economics and economists blend the various philosophies in conducting research.

Economics is a problem-solving, decision-oriented discipline by the nature of its focus. The problems/issues may be disciplinary, subject-matter, or problem-solving in nature. Economics addresses all sorts of problems, ranging from contemporary, short-term problems of everyday activity to complex theoretical disciplinary problems of intricate cause-effect relationships. It is the study of human activities in the allocation of limited resources in satisfying human wants; its focus is in allocating these resources among competing uses. Economic research is expected to be relevant to understanding and dealing with current and future societal problems. This expectation exists both within the discipline and in the minds of our external clientele. How well we fulfill those expectations is a matter on which there will be further discussion.

The three philosophies—logical positivism, objective normativism, and pragmatism—may be relatively more or less important according to the type of research. In disciplinary research, in which we can include basic theoretical research, basic descriptive (including measurement) research, and research on specific analytical procedures that may be applied within the discipline, both logical positivism and objective normativism provide philosophical focus. The part of objective normativism that emphasizes knowledge of values as they relate to behavior is particularly important in theoretical constructions that describe and explain economic behavior of individuals and groups. Logical positivism is an important component of the development or adaptation of basic analytical procedures, but pragmatism is certainly not absent in disciplinary research because often disciplinary research is undertaken with a view toward a social problem. However, problem-solving pragmatic prescriptions are usually less prominent in strict disciplinary research than in subject-matter or problem-solving research.

Subject-matter research pulls from all three philosophies, but in ways different than disciplinary research. In subject-matter areas, for example, environmental economics, labor economics, managerial economics, and natural resource economics (there is no definitive list of subject-matter areas), the objectively normative content often leans more toward development of prescriptive knowledge based on both value and value-free knowledge. The philosophical intent is to develop understanding of outcomes of proposed actions, given specified conditions, in the area of interest. This understanding leads to guidelines and decisions. In any event, prescriptive subject-matter knowledge may be based in part on knowledge of behavior developed through disciplinary research. Subject-matter research also has a pragmatic focus because the prescriptions must achieve the desired effects (they must work). Subject-matter research can be restricted to being only positivistic in terms of its empirical estimation processes and procedures. For example, empirical estimation of the magnitude of impact of pollution user charges on the emission of pollutants or estimation of the amount of an industry's production response to price controls is positivistic.

Problem-solving research that addresses a particular management or policy question for a specific decision maker also can be both positivistic (e.g., in estimating parameters) and normativistic (e.g., in estimating consumer surplus) in its empirical estimation procedures. However, it is conditionally prescriptive (normative) in terms of its objectives (it intends to lead to a decision on what is to be done). Also, it always has pragmatic content. Pragmatism is paramount in problem-solving research because of its emphasis on a particular kind of prescriptive knowledge. While problem-solving research in economics necessarily must pull from conceptually sound and logically applicable theory and employ appropriate analytical techniques, meeting those criteria is not enough; it must achieve the problem-solving objectives of the research. In this regard, Johnson (1986a, pp. 83–84, 86) states:

> Many practicing economists have simply ignored the restrictions of these [philosophical distinctions] as impractical abstractions generated by academic economists out of touch with important problems of the real world. . . . We should note that the complex, sometimes clumsy, methods of pragmatism at least address welfare and value questions in an objective manner, which is more than can be said for positivism and the variants which have emerged from it in economics.

In practice, the three types of research are usually not easily separated. A given research endeavor/project may embody multiple types. Some Ph.D. dissertations, for example, make theoretical or other disciplinary contributions, then proceed on with a subject-matter application and even to some problem-solving prescriptions. Some believe that the discipline would be better served with somewhat more emphasis on empirical analytical validation of theory (Leontief, 1993; Caldwell, 1982; Eichner, 1986; O'Sullivan, 1987, p. 74; Blaug, 1980, ch. 15; Tomek, 1993) as well as the empirical estimation process in subject-matter and problem-solving research.

The position advocated here is that every type of research is relevant, because we need all types. It is quite possible, however, that the collective discipline can be out of balance among the types. If we have, as some maintain, placed too much relative emphasis on disciplinary journal publishing compared to other types of outlets within the discipline, there are two basic forces that may have caused it. First, academic administrators (about two-thirds of Ph.D. economists are employed by academic institutions [Hansen, 1991, p. 1060]) are prone to recognize only journal publishing as evidence of research productivity, coupled with the propensity for economics journals to want to publish only research that is primarily disciplinary in its focus. This system of rewards and constraints has created a situation in which economists, particularly academicians, increasingly communicate only with other economists, and only about things that other users of our research may view as too esoteric to be relevant. Secondly, in their desire to be more scientifically rigorous in a limited positivistic sense, economists have become increasingly enthralled with mathematics, sometimes losing the reason for employing mathematics (to gain more precision in economic concepts) in the rush to express thoughts in mathematical terms and sometimes at the expense of empirical testing and verification. Among other problems (e.g., making assumptions about economic behavior or motivation merely because they facilitate mathematical derivation or deterministic solutions), this fascination may have led us further away from producing applied economic research that can be understood by potential users of our research.

On the other hand, a part of the problem is that most noneconomist users of our research wish for results to be intuitively obvious, with little need for them to understand the complexities of why results occur as they do. This is not possible with many of the complex phenomena that are the objects of economic analysis. A related problem is that very few of our corporate and public policy research users who are interested in applied economic research results understand the need for disciplinary research as the capital stock for applied research. Although solutions to this dilemma are elusive, we need to recognize it because in the final analysis it is our problem; we are the ones who must obtain the support for our discipline and the research within our discipline.

To synthesize how different philosophical positions have merged to affect economic thinking and economic research, Randall (1993, p. 54) states:

> From the rationalists, we have learned that logical coherence is a highly desirable property of an argument, and we have some well-established principles of logic to guide us. From the empiricists we have learned to respect the evidence and to develop procedures to impose *ceteris paribus* and to come to terms with stochastic phenomena. The logical positivists taught us to respect the distinction between empirical and metaphysical propositions, just as the disputes that led to the dismantling of their school taught us that the distinction is not quite as simple as it might seem. From falsification, we learned to cherish the opportunities to conduct a definitive test of an interesting, refutable hypothesis, on the rather rare occasions such opportunities are presented to us.

A similar synthesis, stated another way (Johnson, 1997, pp. 262–263) is:

> Economics has difficulty being either *just* logically positivistic or *only* nor-
> mativistic. If economics is to define and locate optima, it requires both the
> value-free knowledge obtainable with a logically positivistic approach and
> the knowledge of values obtainable with a normative approach; this implies
> either (1) both a logically positivistic and normativistic orientation or (2) a
> pragmatic orientation.

EMPIRICISM IN RESEARCH METHODOLOGY

The influence of the philosophy of logical positivism fostered interest in measurement or quantification—use of data to test validity of theories and derive
expected magnitudes of effects as a basis for policy recommendations. Logical
empiricism may be viewed as going beyond logical positivism to subjecting the
results of logical positivism to testing.[3] There was an interest in quantification of
economic phenomena with the classicists, but quantification remained very limited until the early twentieth century. The simultaneous evolution of several tech-
nologies and events has coincided in the last century to produce emphasis on
empiricism—measurement and quantification—in economic matters.

The interest of governments in the industrialized countries in collecting and
compiling data to produce many types of economic data and indicators has been
associated with positivistic and pragmatic philosophies (a recognized need to
know what it is and to use that knowledge in seeking solutions to perceived problems). This interest in producing social and economic data has both fostered the
development of statistical methods, especially that branch of statistics called sampling, and been fostered by statistical methods, especially the estimation techniques branch of statistics. The emphasis on data, some of it dealing with factual
information about values, has been an important element of empirical developments in economics. The development of the theory and methods of statistical
analysis during the twentieth century has enabled economists to exploit those
data.

As economists learned about statistics, its integration with mathematics and
economic theory led to the development of econometrics. Econometrics represents both a subject-matter specialization and a necessary part of every economist's basic analytical training. Econometrics has its origins in the early 1900s,
but it achieved "legitimacy" in the 1940s and 1950s, made more effective by the
development of electronic computers beginning in the 1940s. Econometrics
includes empirical data and measurements, making it both more positivistic and
more pragmatic and thus more than merely the abstract logic of mathematical
economics (see Johnson, 1986b, p. 84, for further discussion of this distinction).

The influence of applied areas of economics in the evolution of logical
empiricism and the development of econometrics, as well as other empirical

quantitative techniques, has been considerable. Some excellent perspectives and interpretations of these matters are provided in Leontief (1993), Bonnen (1986), Breimyer (1991), Castle (1989), and McCloskey (1990a). In many instances in the evolution of econometric methods, the applied problem-solving focus of the applications led to the questions that identified the need for econometric concepts and theory. A corresponding influence was the role of the institutionalist perspective by the 1930s. The role of institutions was made more obvious by the expanding influence of the U.S. government in public policy matters. The empirical bent (including training in statistical methods) made the contingent of applied economists with sound theoretical knowledge suited for a leadership role in the development of econometrics.

Several leading economists, including Wassily Leontief (1982), a leader in quantitative methods in economics, have pleaded for more empiricism in economics. They look to measurement, reliance on data (objective evidence), and quantitative analysis as the vehicles to place economics on the same scientific footing as, for example, physics.

THE SCIENTIFIC APPROACH

The general process called the scientific approach,[4] which some refer to as the "scientific method," is influenced by the philosophical views previously discussed. We will avoid characterizing the process as the scientific method because there is no single scientific method. There is, however, a central scientific methodology or scientific approach (the two terms may be used interchangeably in this context). It may be presumptuous to label it *the* scientific approach because it is comprised of many variations. The rationalization for doing so is for brevity; if the label is objectionable, substitute the term "a generalized manner of approaching research" or some such descriptive phrase.

The scientific approach may be characterized as having the following general steps: (1) identify the problem/issue/question, which may be disciplinary, subject-matter, and/or problem-solving in nature; (2) define the research objectives; (3) develop approaches for achieving objectives (which may include hypotheses of expected outcomes and/or alternative solutions); (4) conduct the analysis (obtain appropriate information and evaluate it, which may include testing of hypotheses); and (5) interpret the results and draw conclusions, including providing prescriptions, if appropriate. This general list of steps may vary with circumstances. However, if one views the above steps as a guideline for a general progression rather than as an inflexible, lockstep procedure, it is adaptable to most research situations.

The above list of steps is common across disciplines except for step 5. In some fields of research, particularly in the physical sciences, interpreting results and/or drawing conclusions is often omitted. Prescriptions are also less likely to be provided in disciplinary research. Interpretation of results—evaluating the implications of rejecting hypotheses or not—is universally accepted as a part of the process in the social sciences and seems to be increasingly accepted in the physical sciences. Einstein (1991, p. 829) notes that the scientific approach, without step 5, can teach us nothing beyond how facts are related to and conditioned by each other, and intelligent thinking must play a role in the approach. Also, the Committee on Science, Engineering, and Public Policy (1995, p. 2) states that empirical evidence alone, without step 5, is not sufficient. The reader should note the similarities or overlap between this list of steps in the scientific approach and the diagram of steps in the research process in chapter 2 (see figure 2.2) and the relationship between research and science discussed in chapter 3.

The scientific approach is often presumed, especially by persons in the physical sciences, to be based entirely on positivistic philosophy. That is an indefensible position to most thoughtful positivistic, as well as normativistic and pragmatic, economists. Identifying a problem and deciding on research objectives contain elements of normativism and may well contain elements of pragmatism. Problem identification must be affected by individual and group perceptions: that is, what we perceive to constitute a problem. Personal characteristics of mind, individual and public values, and priorities are inherent in the process of problem identification, including disciplinary problems. Likewise, objectives—the identified set of specific goals that are related to the problem—are inherently normative to some extent. A decision must be made about what objectives are appropriate, given the problem. Both problem identification and objectives specification may have a pragmatic orientation. For example, a problem may focus on knowledge that is needed for prescriptions, with the corresponding objectives being designed to provide that knowledge and/or prescriptions. The approach will typically contain more prominent elements of pragmatism when it is focused on problem-solving or subject-matter research than on disciplinary research.

Steps 3 and 4, developing and conducting the analysis, data collection, and testing hypotheses, may be, but are not necessarily, positivistic. In any event, it is through these steps that disciplines are concerned with validation of concepts, as well as hypotheses about results. As has been noted, normativistic knowledge is usually involved in economics and the other social sciences. Recall that the division between logical positivism and objective normativism is somewhat nebulous. These steps, applied for many years in various forms of the classical experimental design in the laboratory and field sciences, are perceived by some to consist entirely of *statistical* data generation and hypothesis testing. As noted in chapter 3, hypotheses and hypothesis testing cannot necessarily be limited to statistical hypotheses. Hypotheses may also be prescriptive (remedial), as noted by Tweeten (1983).

The laboratory and field sciences base these steps of the process more heavily on positivistic philosophy than do the social sciences. The differences in applications of the steps is probably due more to the evolution of disciplines' perceived or defined roles than to any other single factor. The laboratory and field sciences see their research process as producing reliable *data* (e.g., much of their attention is devoted primarily to experimental design to generate statistically valid numbers). The social sciences see their process more in terms of using data to understand *relationships* (e.g., their attention dwells on using numbers to generate conceptually and statistically valid relationships) and to address problems that require decisions.

This latter point is related to the differences among disciplines regarding the role of the last step, interpreting results. Persons whose philosophy is more positivistic may argue that interpretations introduce normativistic elements into the process. Economists concur with that observation but maintain that normative interpretation is a valuable addition if it is done with personal objectivity (neutrality). The data and/or relationships generated or estimated through the process do not necessarily "speak for themselves," as might be maintained by the positivist. They furnish knowledge or understanding, but it is dependent to a degree on the conditions under which it was generated. The scientist's clarification of those conditions can help avoid misinterpretation or misuse of the information. If the dangers of misinterpretation are greater when the scientific approach is applied to more complex relationships in the social sciences, this may explain the differing views among disciplines about appropriateness of interpretation.

In essence, differences in specific applications of the scientific approach revolve around different emphasis on (1) the type of problem (disciplinary, subject-matter, or problem-solving); (2) the nature and types of hypotheses, including explicit vs. implicit and quantitative vs. qualitative; and (3) the roles and functions of implications, conclusions, and/or prescriptions. These variations are not too surprising upon recognition of the different roles and functions in the larger picture of generating new, reliable knowledge of different types. The disagreements seem to stem from the lack of understanding or perspective. They may be unproductive. Different fields and types of research have reasons to place more or less emphasis on the various steps of the approach.

Deduction and Induction in the Scientific Approach

Except in the somewhat specific circumstances when the intent of research is totally on development of theoretical constructions, the approach in economics consists of an ongoing interfacing of deduction and induction. Even when the focus is on developing theory, a deductive process, the theory must eventually be evaluated for applicability or validated through empirical testing, an inductive process. Volumes have been written on deduction and induction and their uses and functions in science, including economics (see, e.g., Larrabee, 1964, pp. 65–91, 236–255; Caldwell, 1982, pp. 36–45; Johnson, 1986b, p. 77). Our treatment of deductive and inductive reasoning is focused on the ways they are used instead of philosophical debates about relative merits.

In deductive reasoning we start from premises—assumptions. Deductive reasoning holds that if individual premises are true and collectively complete and the

4 Philosophical Foundations

reasoning is correct, the conclusion is necessarily reliable. This enables us to organize premises into patterns that provide valid conclusions; however, the conclusions cannot exceed the logical content of the premises (Ghebremedhin and Tweeten, 1994, p. 9).[5] The discipline of mathematics is based largely on deductive reasoning; it is tautological.[6] The most direct application of deductive reasoning within the scientific approach may be in step 3, developing hypotheses. For example, in arriving at hypotheses regarding expected outcomes in economics, we arrive at those expectations through a process of theoretical logic. In disciplinary research the theoretical reasoning may be newly developed; it may entail new concepts, new conceptual elements added to existing theory, or conceptual modification of existing theory, all of which lead to hypothesized outcomes. In research with more applied elements the reasoning will involve the application of existing theory to the issue of concern, with the resulting hypothesis relating to the expected outcome in a specific situation. In the former case, the hypothesis is not only of a different nature, it is more central to the basic intent of the research. In the latter case, economists often neglect to state a formal hypothesis. This is particularly the case with econometric models, for example, and when we are dealing with qualitative hypotheses.

Induction is an empirical process of arriving at new generalities from observed realities (facts, data) and does not depend on previous knowledge (Gebremedhin and Tweeten, 1994, p. 10). Judgments about the generalities can be made based on the probability of making errors. In fact, Churchman (1948) presented statistical inference as a decision theory of science, not as a scientific method. Again, while the process of arriving at hypotheses is largely deductive, the testing of hypotheses, statistical or otherwise, is largely inductive.

As economists we receive formal training in the use of statistical induction. Statistical induction is the process of testing whether estimates of parameters are different from some specified quantity or whether estimates of relationships explain a statistically significant proportion of the variation in variables we are attempting to explain. For the statistical tests to be valid, the population and sample data distributions must meet specific conditions. When we test the statistical significance of, for example, the income coefficient for the demand for clothing, we are attempting to infer inductively something about the impact of income changes on clothing purchases.

Induction cannot constitute proof of a proposition because we have not examined (cannot examine) all possible evidence relevant to the case (Larrabee, 1964, p. 236). Induction is the process of reasoning from specific circumstances to generalized outcomes. We can show that the outcome is probable on the basis of available evidence that has been generated or obtained from reliable information. We cannot say that an outcome or relationship will always hold because it has held in every case that we have examined; the underlying causal structure supporting the outcome or relationship could change (Caldwell, 1982, pp. 40–41). But the outcome based on the evidence is probable, or even has a specified probability.

Although not as highly formalized, the fifth step in the scientific approach is nevertheless largely inductive. When we develop and examine implications of the findings, we are often extrapolating from the specific analysis to the more general applications or prescriptions. Implications may be in the form of policy or management outcomes/prescriptions or in the form of more basic understanding of phenomena or values. As in the more formal inductive process of quantitative hypothesis testing, the specific outcome is not necessarily of primary interest, but it *may* suggest outcomes in the general case. Examples of this in economics lie in the use of Marshall's (1964) representative firm as a vehicle to reason through the expected behavior of the suppliers in an industry. If firms are similar in their cost structures and technology adoption, an increase (decrease) in the price of the good being produced will elicit an output response by individual firms such that the industry supply curve can be viewed as the horizontal summation of the firms' marginal cost curves. This reasoning is sometimes used in empirical research, particularly when the industry is structurally competitive. By modeling the structure/behavior of a representative (i.e., average or typical) firm, the outcome for the industry may be estimated to be the outcome for the firm times the number of firms in the industry. These types of inferences require explanation and defense of the inductive reasoning by researchers on a case-by-case basis.

Identifying the research problem and objectives are both inductive and deductive, with the dominant logic depending in part on the point of departure. If one begins with a vague curiosity about something, the process is likely to be more deductive. Most of us have on occasion begun a research process with "I wonder why (or how, or if) . . . ". When the process is initiated in this manner, one moves from that vague curiosity to more carefully framing the dimensions of the curiosity, then to specifying the intent of the curiosity into a perspective of the issue that lends itself to development of new information, then to research objectives. This is a deductive process of moving from the general (curiosity) to the specific (research objectives).

Alternatively, if one starts with a "felt difficulty," which might take the specific form of a general objective or a researchable problem after being formalized, the thinking process may assume more of an inductive character. If one starts with a specific question or issue ("I would like to determine [or know, or find out] . . . "), then the question of why it is important to get an answer follows in order to justify the objective. This thinking or questioning process moves from the specific (a general objective) to the general (an explanation or understanding of the problem) and is inductive.

Both approaches to problem description are valid and neither is necessarily superior. The deductive approach to identification of problems and objectives may be more common with disciplinary research, while the inductive approach is more the case with applied research. More will be discussed on this matter in chapter 6.

Obtaining data in empirical economic research depends on a deductive reasoning process. Knowing what data are needed is preceded by a process of conceptualizing the relevant variables and often a step involving "model building." Thus, we attempt to deduce the relevant variables on which we need to obtain data. The process of obtaining the data may involve induction if statistical procedures are a part of the process.

Francis Bacon viewed research as a purely inductive process (empiricism) in which we start with facts (what our senses reveal) and proceed "upward" to axioms, then laws (Back, no date). However, the scientific approach integrates the inductive and deductive methods of reasoning. Deduction provides the necessary implications of premises, which may be general (laws, axioms, principles) or specific (factual), while induction examines the validity or applicability of the premises. In isolation, either approach leaves gaps in our understanding. Deductive reasoning can organize what is already known and deduce new relationships, but is not sufficient as a source of new knowledge; inductive reasoning fails to use prior knowledge, (including theory), so it is inefficient.

Deduction (i.e., theorizing) alone is fallible in studying the real world because it lacks definitive means of evaluating whether the premises (assumptions) of the "ideal types" match the situation being analyzed. Induction alone (i.e., observing, empirical testing) is fallible because there is always a probability of error (Ladd, 1991, pp. 4, 5). Put another way, "no theoretical question is answerable without some answers to questions of fact, and vice versa" (Back, no date). Scientific inquiry must rely on the use of both deduction and induction in a constant interaction with one another; we call it the scientific approach.

SUMMARY

Three major philosophies—positivism, normativism, and pragmatism—form the philosophical basis for research methodology in economics. Each philosophy affects economic thinking and economic research, yet no individual philosophy is dominant and no single philosophy sufficiently encompasses the range of economic research. Economics blends the philosophies, as do other disciplines, with the specific mixture varying with the type of research, even the interests of researchers.

Logical positivism as a philosophy related to research emphasizes determination of value-free information and knowledge of values about phenomena and its logical extensions, and its focus is on determining "what is." Logical positivism has influenced economics during the twentieth century, with its impetus facilitated by empirical data availability and inductive techniques and technology. The *relative* influence of logical positivism may have declined since the last one-third of the twentieth century. Objective normativism as a philosophy is imbedded in the foundations of economics. We study how and why people, groups, and societies value things. Economists view objective knowledge of values as being much like positivistic knowledge of physical entities in regard to usefulness, validity, and rigor of the knowledge. Economists also embrace prescriptive normativistic knowledge, conditional on it being personally objective.

Pragmatism as a philosophy in economics is revealed in the emphasis on relevance—our attempts to produce knowledge that is useful in addressing contemporary problems. It does not distinguish between value-free knowledge and knowledge of values because both are interdependent for addressing practical problems. Pragmatism is tempered by overriding considerations for theoretical and analytical rigor, but the identifiable interest in "workability" in our economic research is based on pragmatic beliefs.

The scientific approach—identifying the problem/question/issue, defining objectives, developing hypotheses or alternative solutions, designing research procedures, analyzing information, and interpreting/concluding/prescribing—adapts elements of the three philosophies. Each philosophy may assume different roles and importance in the various components in the scientific approach. Deduction and induction, as forms of reasoning, also have critical roles in the scientific approach. While both are important, with each having functions in different components of the scientific approach, their interaction in the research process allows us to reach implications and conclusions that neither form of reasoning can achieve on its own.

RECOMMENDED READING

Johnson. Chapters 4–9, *Research Methodology for Economists*.
Ladd. "Thoughts on Building an Academic Career," 3–6.
McCloskey. *If You're So Smart*, pp. 1–9.
McCloskey. "The Rhetoric of Economics."

SUGGESTED EXERCISES

1.   Choose a research article from a recent issue of an economics journal. Write a description of how the research reported on was influenced by the philosophies of positivism, normativism, and pragmatism. Comment on how the philosophies have been blended.

2.   Select any recently published report of research. Provide a written description of how deductive and inductive logic were used in conducting the research.

3.   Based on your own reading (knowledge of literature) or research experience, cite an example of research that demonstrates the pragmatic interdependence of value-free knowledge and knowledge of values.

NOTES

1.   The ideas offered on this distinction were initially provided to the author by George W. Ladd.

2.   This should not be confused with conditional normativism, although they are related to one another. *Conditional normativism* takes an individual's or group's values as given without judging

them by subjective criteria; conditional normativism is conditioned on assumptions. *Objective norma-tivism* holds that statements or knowledge about values (good and bad) may be as objective as statements or knowledge of physical things; "both are mental and products of the mind" and both require faith and interpretation (Johnson, 1986b, p. 58). Both objective and conditional normativism prescribe actions or solutions that achieve goals defined by the individual's or group's values.

3.   This perspective was provided by Warren Samuels.

4.   This refers to a general manner of going about obtaining new knowledge or understanding.

5.   Sometimes the premises are true by definition. When this is the case, the conclusions drawn from them are called *tautologies*.

6.   Logic and mathematics are tautological and nonempirical (not testable) and are the only disciplines where certain answers are possible (Simkin, 1993, p. 16). They also have nothing to say about "reality."

II

The Research Project Design

5

Planning the Research

Previous chapters provided general perspective and understanding about research in economics—what research is, what it can and cannot do, its role and purpose in the quest for knowledge or understanding, its relationship to and dependence on science, and how philosophical perspectives influence research activities. This understanding is useful and necessary for those engaged in research activities to comprehend and appreciate how individual research activities and efforts contribute to the general body of reliable knowledge.

But the foregoing chapters do little to provide guidance on how to *do* research. The generalizations about types of research and philosophies, when taken in isolation, probably constitute necessary, but not sufficient, understanding for those attempting to become economic research practitioners. What do I do? Where do I start? These questions and the perceived need to provide useful guidance for students were the reasons that this book was undertaken. This chapter and subsequent chapters are written with a view toward students' perspectives and needs as they endeavor to engage in research. Our system of higher education tends to "feed" students their training and understanding (the mixture of training and understanding varies) in somewhat self-contained subject-matter packages. The system appears to be effective as evidenced by the United States being recognized throughout the world for its superiority in graduate education. However, for students to make the transition from absorbing what is already known to using what is already known to produce *more* information or knowledge, we leave them to their own devices except for the individualized assistance (apprenticeship) from their teachers, graduate advisors, and committees. The contention is that both the students and their advisors can benefit from some prescriptive guidelines that help to provide a structure, yet are flexible enough to facilitate individual creativity and the circumstances and needs of particular research efforts. Again, this and subsequent chapters are heavily focused on the perceived needs of students. In the larger sense, we are all students; some have just been at it longer than others.

The material in this part of the book is not particularly unique or insightful when considered with hindsight of experience. It is time tested and very dependable. It is deceptively simple as a concept or approach, yet inherently difficult and taxing in implementing as a specific project. As a general approach to the process, it will rarely, if ever, fail. It is *the* approach to problem (or issue or question) solving in economic science, and to a large extent works for problem solving of any kind, be the problem disciplinary, subject-matter, or problem-solving. That is, the approach of identifying and understanding the problem and objectives, then designing an approach to accomplish those objectives is a reliable formula for success with all kinds of problems—research, policy, management, personal, and so on.

This part of the book—chapters 5 through 9—will make frequent references to appendices A, B, and C. These appendices consist of example research proposals. Their purpose is to help illustrate the principles and guidelines prescribed in the procedural part of the book. The research proposal in appendix A was selected to represent a proposal prepared for a government agency. The proposal in appendix B is a graduate student thesis proposal for an M.A. program in economics. The proposal in appendix C is for a Ph.D. dissertation. These proposals will be referenced to illustrate, and sometimes contrast, certain procedural points. The proposals as a group help to show, among other things, both similarities and differences in structure, approach, and sophistication in research proposals.

The first section of this chapter discusses the research proposal as the planning vehicle, followed by a section that describes the common components of research project proposals. The third section covers some of the considerations in evaluating proposals, followed by a brief discussion of effective use of the proposal and some perspective of public support for economic research. The final two sections address the importance of writing in planning, conducting, and reporting research and some guidelines on scientific and technical writing in economic research.

THE RESEARCH PROJECT PROPOSAL

The key to successful research is the *research plan*. Recall that a central element of the definition of research is that it is a systematic approach (chapter 2). In order to target understanding or knowledge on specific matters, be they disciplinary, subject-matter, or problem-solving oriented, the approach needs to be carefully constructed and designed to ensure that the information or knowledge produced by the activity is reliable. To derive the knowledge without a design occurs only by accident. This does not say that it is impossible to discover public reliable knowledge by accident: that is, by simply pursuing idle curiosity. But accidental discovery happens in a small proportion of cases and is not dependable—like relying on a random event to deliver the understanding. The heart of the research plan is the *research project proposal*.

The intent is not to denigrate the importance of accidental discovery. Being open to new thoughts, ideas, approaches, and insights is an integral part of creativity in the research process. But the most insightful accidental discovery in economic research usually occurs within the boundaries of a structured direction of inquiry. The research proposal should not limit creativity, but rather channel it toward objectives the researcher has discerned through a reasoning process.

Research proposals are generally required for any type of research endeavor and by all entities that support research. Specific requirements for proposals vary drastically. Research proposals that seek financial support differ from research proposals that serve only as the guide for the researcher. The most stringent requirements in terms of completeness and exactness may be for Ph.D. dissertation proposals, although some federal agencies, or some types of federal research funding programs, require elaborate, more highly planned and detailed proposals (see appendix A). Other funding sources, for example, some industry groups, may prefer proposals that are concise, brief, and to the point; they may wish to forego conceptual and other "academic" aspects of the research in the proposal. In rare cases the proposals may be oral instead of written, although when oral proposals are required, they usually are in addition to a written proposal.

Irrespective of the specific requirements of the proposal, it constitutes evidence of the research plan. The more thorough and clear the proposal, the clearer and more complete the plan. The importance of effective written communication skills is difficult to overstate, not that communication skills will always produce good proposals, but they are impossible to produce without communication. There must first be clear, well-developed thoughts that are then communicated effectively. To whatever extent *both* are not present, the proposal will fail to achieve its purpose. When the proposal is not clear, the only reasonable assumption the reader can make is that the writer's thinking is not well developed. When someone says, "I understand it, I just don't know how to say it," it really means, "I don't understand it clearly enough to explain it."

The proposal thus serves dual purposes, one for the researcher and one for the evaluator(s) of the proposed research. For the researcher, it provides an operational plan for conducting the work while simultaneously keeping the researcher from wandering off on questions or issues that are, albeit worthy, not germane to the purpose of the study. It is also a vehicle that forces the researcher to more thoroughly understand the reasons for and intent of the research, to anticipate potential problems, and develop plans for dealing with them before they occur. For evaluators (e.g., administrators, funding individuals or groups, or graduate student advisory committees), the proposal is the means through which they determine the intent of the research and make approval/disapproval and funding decisions. It also provides the means for them to require accountability from the researcher.

ELEMENTS OF THE RESEARCH PROPOSAL

Although proposals vary in their complexity and configuration, there is a set of common parts or elements. These are identified and discussed briefly below. The listing is not exhaustive, but it is representative, and all proposals do not contain all elements. Figure 5.1 is used in conjunction with the following listing to help

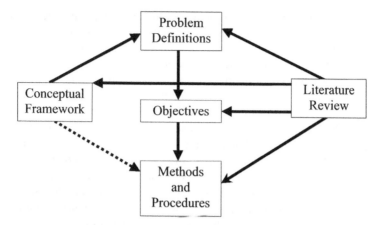

Figure 5.1. Linkages among components of a research project.

describe how some of the main working components of the proposal relate to other components. The list draws on the material of Williams (1984).

A. *Title*. The title of the project should be descriptive of the major focus of the research, but it should be no longer than necessary. The title cannot possibly convey a thorough understanding of details of the study, but it can provide an accurate impression of the central focus of it. Doing so in the most efficient manner (the least unnecessary words) is desirable. For example, a project title such as "An Economic Study of the Impacts of Lowering Import Tariffs on Textiles on Consumers, Textile Manufacturers, and Fiber Producers in the United States" can easily be shortened to "Lowering Import Tariffs on U.S. Textiles: Effects on Consumers and Industries." Both titles convey approximately the same understanding of the intent and content of the proposed research, but the latter does so with fewer words.

B. *Identifying information*. This tells something about the people and organizations that would be involved in the proposed research and other types of summary information relevant to the project. For example, the names, titles, addresses, and telephone numbers of project personnel may be given. The people may be divided into groups such as project leader (or principal investigator, etc.), research advisors or cooperators (individuals or organizations) and, in the case of graduate student proposals, graduate committee members (see appendices B and C). An indication of the budget for the research and specification of the time required for the research (dates for project initiation and completion) may be included. Additionally, a detailed budget, when needed, is usually presented elsewhere in the proposal. A detailed work plan may be included elsewhere, as well as biographical résumés of the research personnel. When résumés are provided, they are usually appended to the proposal rather than integrated into the body of the proposal. The type, amount, and format of the identifying information may vary and should be developed with (1) instructions about format specifications and (2) the readers of the proposal in mind. Some examples of formats for identifying information are shown in appendices A through C, but there are many acceptable variations.

C. *Problem identification and explanation.* This part of the proposal describes the problem being addressed by the proposed research and provides the rationale for the proposed objectives (represented by the solid line between objectives and problem definition in figure 5.1). The problem may address disciplinary, subject-matter, or problem-solving issues. It is perhaps best described in a two-step procedure, with the first step developing the general perspective of a broad problem area and the second step narrowing attention to a carefully developed subpart of the general problem that is researchable within the resource constraints of a single project. The researchable problem is the reason (justification) for the proposed research. In some cases this section of a proposal may even be titled "Justification." In other cases, format may call for both a problem statement section and a separate justification section, but when that is the case the two sections are closely related. A problem statement or justification is, to the author's knowledge, a universal requirement for research project proposals. Chapter 6 takes up the matter of problem definition in further detail.

D. *Objectives.* The objectives, also universally required in research proposals, specify precisely what the proposed research project intends to find out, discover, and/or accomplish. Objectives identify the goals or ends of the research, not the means. Objectives are usually best stated as a one-sentence general objective and a list of specific objectives. They need to be stated clearly, but concisely, and oriented to producing or discovering knowledge or information, or perhaps prescriptions in subject-matter or problem-solving research. Objectives are justified by the problem statement and are the point of focus of the methods and procedures (figure 5.1). Further elaboration of these principles is also provided in chapter 6.

E. *Review of literature.* This part of the proposal provides a summary review of the research literature that is relevant to the study being proposed. Its nature and structure can vary. For example, it may be a separate section in the proposal or it may be integrated into other sections (problem, methods and procedures, or conceptual framework). The literature reviewed may be related to the proposed research through the problem, objectives, methods and procedures, and/or conceptual framework, as indicated by the solid lines to those components in figure 5.1, but it should be confined to scientific, as opposed to popular, literature. Its purpose is to provide the base of knowledge on what we have already learned about matters that are addressed in the proposed research. This knowledge base is for the benefit of both the researchers and the evaluators of the proposed research. Although the literature review is presented here after the objectives, in practice it is difficult, perhaps impossible, to fully develop the problem statement and objectives prior to conducting a thorough literature review (although the written literature review does not necessarily have to be written at that point). All proposals do not require a formal literature review section, but no research project should be undertaken without the researchers being knowledgeable of the prior research related to the topic. The review of literature is the topic of chapter 7.

F. *Conceptual framework.* A formal conceptual framework is not a universal requirement for research proposals, but any economist who attempts to forego this step risks making potentially serious logical errors in his or her research. A formal conceptual framework should be standard in research proposals by students, if only to ensure their comprehension of the principles as they relate to the research problem. In its most basic form, the conceptual framework is an analysis, using economic, and perhaps other, concepts of the specific researchable problem that the proposed research addresses. *It is a conceptual analysis of the problem.* Its relationship to other parts of the proposal is illustrated in figure 5.1. It may be directly affected by the literature review. The conceptual framework also may lay the foundation for some of the research methods and procedures (the dashed line in figure 5.1), but that is a side benefit, not its central purpose. The central purpose is to ensure that the researchers and evaluators are examining the problem with appropriate concepts, thus helping to obtain a clear perspective of the research problem/issue/question. In disciplinary proposals for research that is primarily theoretical, the conceptual framework might be very brief, or even absent. In some cases the primary objective of the research may be to develop a conceptual framework. In disciplinary research proposals in which the problem is measurement, the conceptual framework might consist of a theoretical analysis of that measurement problem. In research with a problem-solving focus, the conceptual analysis would be of that type of problem. The conceptual framework is the subject of chapter 8.

G. *Methods and procedures.* The methods and procedures describe how the objectives will be achieved. The methods and procedures flow directly from the objectives, are affected directly by the literature review, and may be indirectly affected by the conceptual framework (figure 5.1). They address matters such as data to be generated or collected, analytical techniques to be used, sequencing of procedures, and derivations from empirical estimates. Inclusion of methods and procedures is universal in economics research proposals, although the detail and complexity may vary to fit the background and preferences of evaluators. For example, a proposal being reviewed by economists probably would need more detail and economic sophistication in its methods and procedures than a research proposal being reviewed by noneconomists. The methods and procedures in student research proposals should contain as much detail as the student can envision in the planning stage of the research project. Methods and procedures are addressed in detail in chapter 9.

H. *References.* The list of references provides the documentation of the sources used in the proposal. References include those sources that may have been used in explaining, defining, or documenting the problem, those used in the review of literature, and any references used in the conceptual framework or the methods and procedures. The references and referencing are covered in chapter 7, in conjunction with the review of literature.

EVALUATING RESEARCH PROPOSALS

When a research proposal is assessed, the soundness of the research plan or design is being evaluated. The proposal is evidence of the quality of thought that has gone into the project, and the whole research plan (i.e., the desirability of the proposed research project) will be judged by the quality of the project proposal. This necessitates clear, incisive thinking and planning, and it needs skillful communication. The planning of the research and the writing of it (in short, the proposal) are usually the more difficult parts of the research task. The process of carrying through with the research plan becomes more a matter of "going through the steps." This is obviously a simplification because thought, judgment, and decisions also will have to be used in carrying through with the plan. However, on a comparative basis, the defensibility of the research rests on the plan. Well-executed procedures on a poor research plan still produce poor research. Evaluators of potential or proposed research projects are concerned with the plan.

In a presentation at a research conference, an administrator (Smith, 1983) provided a list of the criteria by which administrators of research activities and funding evaluate applied (nondisciplinary) research projects. The criteria are not specific to economic projects, but they generally show what administrators and other evaluators of applied research proposals consider. Smith's slightly modified list is provided below.

A.   Is the investigator interested in the problem? This is important because genuine interest provides motivation and fosters creativity.

B.   Is there a genuine lack of knowledge related to the problem?

C.   Is the research recognized as needed by people other than the investigators? The perceived need should extend beyond the disciplinary interests of the researchers to "users" of the research.

D.   Are the objectives:

   1.   appropriate to the problem?
   2.   attainable?
   3.   observable or measurable?
   4.   sufficiently specific?

E.   Does the investigator have sufficient resources (education, training, facilities, material, and funding) for the project?

F.   Does the proposal recognize appropriate constraints, both internal (i.e., constraints of the mission or role of the organization) and external (e.g., constraints imposed by social, economic, or political dictates)?

G.   Does the proposal, as evidenced by background investigation, indicate that the research is likely to be productive?

H.   Is the expected value of the research greater than the cost of obtaining results?

I.   Has the study been designed in a manner such that results are as widely applicable (to as many people, situations, locations, etc.) as is feasible?

These or similar criteria are widely used by administrators who evaluate proposals for problem-solving and subject-matter research. Disciplinary research proposals require a modified set of criteria, but disciplinary research often does not go through a formal proposal review phase. When disciplinary research is proposed and reviewed, it is disciplinary research of known relevance (see Johnson, 1986b, ch. 10 and pp. 195–198 for elaboration on these points). However, the support for disciplinary research is limited, and most disciplinary research in economics is done without overt support or direct administration. These contentions are supported further in a subsequent section of this chapter.

IMPORTANCE OF FLEXIBILITY

While a primary function of the proposal is to provide boundaries and guidance for the research, it must not be viewed as a "road map" from which there can be no deviations. The plan may have to be altered as the research progresses. Occasional modification of an objective may be needed to improve the focus and usefulness of the project. Sometimes one discovers, or even develops, a new method or procedure that achieves the objectives more effectively than the method or procedure originally proposed. One often develops a more thorough perspective of the research problem and/or the conceptual framework as one learns more of the issues while carrying out the research. One often discovers more related literature than was found in the original literature search; this alone can sometimes cause other parts of the proposal to be modified. Proposal writing is an evolving process.

The important point to be made is that the project proposal is an asset in the research process, but it must not become a straightjacket. Do not view it as something that confines your research, but as something that keeps your attention and research efforts focused on the problem and objectives defined for the research. It keeps you from letting your attention stray too far from the point of the research. You want to be efficient in conducting your research, but you do not want to stifle creativity in the process. As you have new thoughts and perspectives and encounter new ideas, consider and evaluate them. Integrate them into your plan if they are warranted and feasible within the general objectives and resource constraints.

A guideline on making proposal revisions after a proposal has already been approved is that minor revisions may be introduced into the plan by the researcher without formal approval, but major revisions should seek approval from the reviewers or approvers of the original proposal. This holds in any situation in which the proposal goes through external evaluation, such as proposals for funding and graduate research proposals.[1] For example, a research project funded by an agency or group would need approval of substantive modifications in any of its objectives but might only require notifying the agency or group of some changes in procedures. Graduate student research could need formal graduate committee approval if major changes in methods and procedures are proposed. Determination of whether formal approval is needed for a proposed change is obviously a matter of judgment. One guideline is always reliable: When in doubt, *ask*.

FUNDING FOR ECONOMIC RESEARCH

Unlike the case with the "hard" sciences, the public perceives little direct need for disciplinary research in economics. Consequently, there is little direct support for disciplinary research in economics. Users of economic research are willing to provide financial support for subject-matter and problem-solving economic research because they can visualize direct benefits from it. Economists are sometimes perplexed by the absence of recognized need for the basic research in economics when they see public institutions (government, universities, foundations, etc.) content to finance disciplinary research in chemistry or biology, for example, that has no known immediate application, but see no need for disciplinary research in economics.

Economists may have done too little to "sell" the need for disciplinary research in economics to the public, the business community, and government. Some disciplines have been able, by accident or design, to create the general understanding that their pool of basic knowledge must be kept growing for benefits to society to accrue. For example, the public seems to accept the premise that basic research in chemistry is necessary; the public is willing to more readily assume or accept that the knowledge gained will eventually provide benefits to the public at large through yet-to-be-discovered applications. Yet there seems to be no such general presumption about economics or any of the other social sciences. There may be more popular support for purely disciplinary activity in the arts than in the social sciences.

Disciplinary research in economics is supported to a small extent by public institutions such as the National Science Foundation, by some government units, by universities that allow or encourage faculty to pursue their own disciplinary interests, and by subject-matter and problem-solving research projects within which some disciplinary research is undertaken as a means of achieving the applied research objectives. We have no reliable information regarding the proportion of support from these types of sources, but their relative importance is probably in reverse order of the above listing. The point is that even disciplinary research in economics tends to be in response to perceived needs to address current issues and problems. The disciplinary research in economics is typically directed to the inability of economic theory or procedures to provide adequate insight and guidelines for issues of the day. John Maynard Keynes's *General Theory of Employment, Interest, and Money* was in response to the failure of the formal economic theory of the 1920s to provide guidelines on how to deal with the problems of a depression economy. Johnson (1986b, pp. 121–138) provides an analysis of two cases of economic research that also supports this position—Wassily Leontief's research on input/output relationships and T. W. Schultz's work on human capital.

There is, however, public support for applied problem-solving and subject-matter economic research. Public administrators, elected officials, business leaders, and even a portion of the general public grasp the need to understand and predict consequences of economic actions and phenomena and economic effects of social and political actions, to generate prescriptive solutions to management and policy questions, and such matters. These groups are interested in positivistic knowledge, normativistic knowledge of values, and conditionally prescriptive knowledge. However, the mind-set is that they want the research to generate the knowledge to be directed to specific, defined problems as opposed to general, conceptual probing for knowledge. Our external clientele take a highly pragmatic interest in our research.

Economists have concerns about public support, financial and other, for economic research. Readers interested in the author's perspective on research funding in economics, including some suggestions for young professional economists, may want to read appendix E.

An offshoot of this discussion is that economists cannot be easily grouped into theoretical economists on the one hand and applied economists on the other. The *relative* emphasis we place on theoretical or conceptual components of the applications may, and do, differ, but we all focus on current issues with which public and private decision makers are concerned. Furthermore, economists who are conducting multidisciplinary subject-matter and problem-solving research cannot conduct that research without also being capable theoreticians; they must grasp the theoretical aspects of the problems they address. However, with applied (multidisciplinary) research, understanding the problem theoretically is not sufficient. Conducting applied research requires the economist to go beyond the confines of the theory (recall that all theory is to some degree a generalization and simplification of the real world). Applied economic research requires that we use the theory to provide a framework for analysis, but the approach also must recognize and deal with other circumstances that must be accounted for and that may modify the theoretical outcome.

IMPORTANCE OF WRITING

"To write well is to think well" (Ghebremedhin and Tweeten, 1994, p. 91). Even if good writing is not, in itself, good thinking, it is the *evidence* of good thinking. It has also been called "visible thinking" (Mighell and Lane, 1973, p. 15). There is a degree of alarm within educational circles, at least within the United States, about the capacity of people in general to write well. The reasons are varied and only vaguely identified, but it has serious implications for research in general and economic research in particular. It is recognized as a problem in graduate programs in economics. Academic and nonacademic employers of economists with new graduate degrees, as well as journal editors and readers, have been vocal in their complaints about their level of writing and other communication abilities, especially their basic expository skills (Krueger et al., 1991, pp. 1048–1049). Even professional economists (e.g., Freedman, 1993) lament that economists (as well as other science groups) generally fail to communicate well enough to be understood because of disciplinary chauvinism, excessive use of jargon, including mathematics, and a more general failure to write well (clearly).

Writing is a skill that can be learned and taught. Some people obviously find writing easier than others, but few find technical or scientific writing an easy task. This may be largely due to the complex nature of the subject matter and the need to be very exact—leaving as little chance for misinterpretation as possible—in the communication. Yet economists should be at least adequate to the task because they must do a substantial amount of reading, which itself fosters the capacity to write. An essential part of developing the skill is to practice writing and to develop a productive, realistic attitude about it. Writing is not an activity that one decides to do, sets down and does it, and is finished. It is a *process*, and the process involves *rewriting*, usually numerous times. Effective communication is important to the research process in several ways, and investment in the time to practice, study, and develop the skill is both productive and necessary. But it is also often demanding, sometimes frustrating, and occasionally rewarding. The author sometimes, in an attempt at humor, tells students, when they are particularly frustrated with the demands of writing, to be encouraged—writing papers gets much easier after you have written two or three hundred of them. It is not always taken with the humor intended.

The importance of writing goes well beyond the function of reporting research, although that is one aspect of its relevance. Writing is the evidence of research productivity. Writing makes research results accessible to others. Writing affects one's professional standing. The writing activity may be more critical for economics than for some other disciplines. But writing is a part of the research process itself (Ladd, 1987, pp. 77–78; Mighell and Lane, 1973, pp. 15–16).

In the process of developing effective written communication, we force ourselves to clarify and refine our thoughts, to place them in logical order, and to express them in ways that enable others to grasp our thought process and reasoning. Few of us fully understand the intricacies of our own thoughts on a matter until we have gone through the disciplined process of explaining them clearly in written form. That is why the *written* research proposal is of premier importance in planning effective economic research. Ladd (1987, p. 79) further points out that writing plays different roles at different stages of the research process. For example, writing in very early stages may entail little more than scribbling notes but require painstaking precision and organization in preparation of final reports. For students, implementation of any of the planned methods and procedures should not be done before a written research proposal is completed and successfully defended.

WRITING GUIDELINES AND TIPS

The purpose of this section is to provide some general insights, suggestions, and admonishments about technical or scientific writing as it applies to economic research. The reader should note that these apply to all technical writing (e.g., research reports, journal articles), not just proposals. Note that this book is not written in technical style; the intent of this book is different from the purpose of a technical proposal or research report. None of the material in this section is unique to any particular source or author. Different aspects of this material can be found in various sources. Two specific references that are both useful and concise are Ghebremedhin and Tweeten (1994, pp. 90–100) and Mighell and Lane (1973, pp. 16–20).

The goal of writing in research is to convey information. Good writing is writing that serves its intended purpose. Research writing does not have to be dull (it should hold readers' interest), but the research audience is seeking information, insight, knowledge, and/or mental and intellectual stimulation rather than entertainment as recreation or diversion. You must assume that your audience wishes to understand and be informed as effectively and efficiently as possible, while not being bored or lulled to sleep in the process.

One recommended element in conveying information in research writing is to determine the intended audience and their information and knowledge needs. Knowing your audience includes an understanding of readers' levels of technical knowledge as well as their interests. A research proposal or report that is to be read by a group of government or industry policy makers must be written differently, with different aspects of the concepts, procedures, and results emphasized, than a report on the same research presented to a group of economists. Case in point: Economics journal articles are written to be read by economists. Hardly anyone except professional economists will even attempt to read a disciplinary or subject-matter economics journal. When noneconomists (and some economists, too) attempt it, they tend to be confused and "turned off" by the experience. A useful technique for some writers is to try to place themselves mentally in the anticipated readers' position as a means of focusing on their interests and prior knowledge.

Another reliable guideline is to *organize* what you have to say. You are trying to achieve a logical order of presentation of the thoughts and information that is clear, direct, and precise. You want a logical sequencing of thoughts, ideas, and information; you are "directing" your readers' thinking and understanding. Achieving this logical order invariably requires an outline of what you are to present. The outline is the tool for organizing the flow—major points, subpoints, linking points to one another. It is advisable to construct a written outline (the more detailed the better) rather than trust an outline to memory.

Communication of logical order of thought and information is fostered by several characteristics that can improve your research writing. One is simplicity. Simple, direct statements communicate more effectively than do complex, rambling, or wordy statements. Use simple statements and phrases and avoid unnecessary qualifiers, jargon, clichés, and platitudes. For example, the sentence "While the estimates of elasticity of exports with respect to export prices made by various researchers show a large range of variation, that being from -0.3 to -1.2, the predominance of the estimates falls within the range of -0.5 to -0.8." might be condensed to, "Prior export price elasticity estimates range from -0.3 to -1.2, but are mostly in the narrower range of -0.5 to -0.8." Of course, the final determination must be consistent with the context.

Another desirable characteristic is precision (lack of ambiguity) in language. Sometimes a minor difference in the choice of words can alter the readers' perception of the meaning of a statement. How often have you seen a writer, occasionally even a professional economist, use a phrase such as "demand increased" when he or she really meant "consumption increased" or "quantity bought increased?" "Demand increased" describes a shift in a demand relationship, whereas the other phrases signify either a shift in a demand relationship or a movement along a demand relationship. The ambiguity or lack of precision on such a fundamental distinction is a grave error for an economist. While ambiguity detracts from communication, we should recognize that simplicity and precision may conflict on occasion and necessitate a trade-off. This may be a problem, especially when the audience contains noneconomists. Technical terms with specific meanings sometimes must be foregone to achieve communication with noneconomist readers. The best advice on these situations is that they require judgments on a case-by-case basis.

Another guideline is to avoid advocacy and judgmental words and phrases. Keep the tone of the writing as impartial as possible. You are not an advocate for any group or position, and it is desirable to avoid any such connotation or appearance. Avoid personally judgmental words (e.g., *should*, *good*, *bad*, etc.). Beware of adjectives such as "unfortunately" or "regrettably" in descriptive phrases. Do not use "poverty-stricken people" when you can be both more precise and impartial with a phrase such as "families with incomes below the government-defined poverty line of $4,000 per capita per year." In short, be as complete, accurate, and objective in your writing attitude as you can.

An essential component of good scientific writing is having what you write reviewed, evaluated, and critiqued by persons who have experience and perspective in the same type of writing. For students, this should include evaluations and assistance from professors. None of us can be as unattached and analytical with our own writing as with someone else's writing. You want reviewers who are critical but constructive, so as to help you identify the weaknesses (and strengths) in the drafts you write. Remember that the service others provide in reviewing your written material is likely to require reciprocation. That reciprocity may, however, also benefit your own writing ability, as indicated below.

A practice that will help you become a more effective research writer is to critique other people's written material. By studying others' written material with a view toward being critical in a constructive way,[2] you develop a greater ability to be more objective and critical toward your own writing. For that reason, requiring students to critique each other's drafts of parts of their proposals and research reports (e.g., theses, dissertations, term papers, articles being prepared for submission to journals, etc.) is a useful practice. Written guidelines for doing critiques are provided in appendix D. Another useful practice that fosters effective writing is to reevaluate the writing after it has been set aside for a time. Write a draft, then return to it later, sometimes days or weeks later, with a fresh perspective.

"Style" is an important part of good writing. Each of us has our own style of writing and there is no single appropriate style. We also must recognize that writing style requirements may vary with the type of publication; always check on any mandated style requirements for any type of outlet. Most journals have some individual style preferences, different government agencies have specific style dictates, and most graduate schools have style guidelines for theses and dissertations. There are also general writing guides, such as *The Little, Brown Handbook* (Fowler, 1983) and *The Craft of Scientific Writing* (Alley, 1987), that provide guidance on matters ranging from punctuation and grammar rules to creating the mood for writing. Another type of useful aid is computer word processing editing software. In developing your own style, it helps to read good writers; it may help you to identify what you are comfortable with and what you like.

There are, however, some general style guidelines irrespective of your individual style preferences. These include avoiding complex phrases and sentences, making sure you have correct spelling (a much easier task with spell check options in word processing software) and punctuation, and avoiding digressions into topics not essential to the purpose at hand. A style technique that is useful in practically all cases is to begin by telling readers what you are going to do and summarize it for them at the end. There are also some technical points of style that may vary, but you may need to be aware of them. These include:

A. *Page margins and line spacings.* These vary with the outlet; unless specified, 1.25-inch top, bottom, and side page margins and double spaced lines are usually safe.

B. *Mathematical and statistical notation.* Some outlets promote technical symbols and notation while others prefer to avoid them (this is a consideration independent of the audience to which the writing is directed).

C. *Citations and reference notes.* Some outlets allow footnotes while others wish to avoid them. Endnotes are an option. Styles of referencing may vary (more on this in chapter 7). Whatever style you adopt, be consistent in its use throughout a written document.

D. *Tables, charts, and graphs.* "A good picture may be worth 10,000 words but a bad picture is worth hardly anything" (Mighell and Lane, 1973, p. 18). Style dictates vary greatly. Most table formats specify no vertical rules and the placement of table titles above the body of the tables. Standard format for charts is for the title at the top as well. Figure titles are standard at the bottom of the graph. However, many current graphics software packages force titles at the top, so this could change as a matter of expediency. A generally accepted rule of tables, charts, and graphs is that they should "stand alone." Titles should be explanatory; they should be presented such that readers can read and understand them independent of the text. All units should be specified. All columns, axes, and so on should be clearly labeled. Footnotes may be used to explain needed details or caveats. Complete reference citations should be provided with the visual if references are needed for sources or other purposes.

E. *Verb tense.* When to use present, past, or future tense can be confusing. In general, be wary of future tense except when writing the methods and procedures section of a research proposal. Preferences vary between present and past tense on proposals and reports. The most reliable rule of thumb, however, is to try to determine the preference of the targeted audience. It is generally better not to shift verb tense within a major section of a written document, although switching between major sections may be warranted.

F. *Personal pronoun form.* In formal technical writing, the rule historically has been to write in third person (e.g., he, she, it, they), not in first or second person (e.g., I, we, you). Journals in particular in recent years are deviating from this form. A safe general guideline is to use third person for formal research reporting, but first or second person is appropriate for less formal style such as presentations and proceedings papers. Consult specific guidelines.

Writing is, to a degree, a personal matter with most of us. We want our writing, as well as our oral communication, to be well received. That "need" is perhaps fostered, even instilled, by our early studies of expository writing; our literature and exposition teachers value "inspirational," "moving," "persuasive" writing that stirs the emotions. Some of these attributes are also valued in technical and scientific writing, but precision of meaning, accuracy, and analytical rigor are more important attributes.

Our consideration of writing and its role in research is not finished. The matter will be a point of attention at several junctures in subsequent chapters.

SUMMARY

This is the first of the book's six procedural chapters. The primary focus of the chapter is an overview of the research plan, embodied in the research proposal, with additional consideration of two closely related matters—planning and justifying research based on proposals and communicating in writing.

The written research proposal constitutes the research plan. Research project proposals may differ substantially with the type and purpose of research and the audience to which the proposal is directed. In general, the proposal is necessary for both the researcher (to organize the research activity and/or obtain support, financial and otherwise, for the research) and the evaluators charged with evaluating, approving, funding, and/or supervising the research activity. There are common elements for research project proposals—title, identifying information, problem description, specification of objectives, literature review, conceptual framework, methods and procedures, and references. All elements are not necessarily required in every proposal and the manner they are included may vary, but they constitute the standard set of elements.

Guidelines are provided on evaluation criteria and on modifying the research plan. The different ways in which economic research is funded are discussed and attitudes and support for the different types of economic research are considered.

Written communication within the context of both proposal writing and scientific writing in general is emphasized. Writing represents evidence of the thought process. It should be recognized as a process, involving rewriting, and as an integral part of the research process rather than an addendum to it. Some general suggestions for effective writing are provided, recognizing that there is substantial flexibility for individual style within scientific writing. Reviewing and critiquing are advocated as means of accomplishing effective writing.

RECOMMENDED READING

Alley. Chapters 1–13, 18–21, and the appendix, *The Craft of Scientific Writing.*
Ghebremedhin and Tweeten. *Research Methods and Communication in the Social Sciences.*
Ladd. *Imagination in Research*, pp. 77–81.

SUGGESTED EXERCISES

1.   Locate three economic research proposals, one written by a graduate student for graduate program research, one submitted to a government agency for funding, and one to an industry group for funding. Describe how and to what degree the proposals differ in their components, detail, and organization.

2. Provide a written constructive critique of the graduate student proposal from the previous exercise.

3. Write a 1–2-page description of the differences between technical and expository writing.

NOTES

1. When the proposal's only function is to formulate the plan for the researcher (i.e., there is no external review), there is obviously no need for external approval for changes.

2. Being critically constructive may consist of a two-step procedure: (1) a critical reading to identify what is right/wrong, clear/confusing, effective/ineffective, and the like; then (2) correcting or making suggestions for changes.

6

The Research Problem
and Objectives

A problem well-stated is a problem half solved.
John Dewey

A problem is a perceived difficulty—a difference between what is (or perceived to be or exist) and what might be, which is inherently normativistic. A research problem[1] is a discrepancy between what is known or understood and what is needed or desired to be known or understood. The discrepancy (problem) may concern disciplinary knowledge, subject-matter knowledge, and/or problem-solving knowledge. Research problems are different from, for example, decision or action problems. Research problems address the need to know, perhaps prescribe, whereas decision problems address the need to decide something or take an action. For example, public policy makers face the need to decide among alternative public policies about many things. In other words, they face policy problems, and business people and other managers face business and management problems that require decisions. We all face decision problems, and we all face action problems. We have problems of choosing among alternative courses of action or things to do. Research may provide knowledge and offer prescriptions, but it cannot make a decision and implement it.

We may view problems and problem solving as being of three interrelated types. Once we know what the overall problem is, it may be segregated into three types of problems, or three steps in problem solving. First, there is a problem of obtaining information that is relevant to the overall problem and determining what that information reveals regarding the overall problem and its potential solutions. In its most basic form, this is the research problem component of the overall problem. Second, there is the problem of examining the alternative approaches to the overall problem and choosing among these approaches. This is the decision problem component, and it often relies on results of studying the research problem component. The third problem component is the action problem. Once the decision is made, then how to implement the decision (take action) must be determined.

Our concern in research methodology lies specifically with research problems. We do not take the position that the other types of problems are not important. Our need to recognize the other types of problems arises from the fact that the research problems, issues, and activities do not exist completely separate from the decision and action problems. Conversely, the need exists to maintain a clear distinction between research problems and the other types to avoid the research activity addressing nonresearch matters. This distinction arises at several points in this chapter.

Various classifications of problems exist. For example, Northrop (1959) identifies four kinds of intellectual problems, those of (1) logical consistency; (2) empirical truth; (3) fact; and (4) value.[2] Back (no date) classifies *research* problems as problems of either intellectual or practical difficulties, suggesting that both disciplinary and applied research problems are important. It also may be noted that meaningful research problems (i.e., those that address an answerable question) and relevant research problems (i.e., those for which the answers are crucial to resolving intellectual or practical difficulties) are related but not the same. For example, some problems may be of intellectual curiosity and capable of being addressed in a research context (meaningful), but have no immediate relevancy to solution of a decision or action problem. This situation occurs more often in disciplinary research than in applied research.

The remainder of this chapter is organized as follows. A distinction is made between research and decision and action problems, followed by a brief consideration of the importance of the problem specification. The researchable problem is then differentiated from the problematic situation, then the problem statement as a part of the research proposal is considered. The use or role of data in the problem specification and a discussion of objectives are presented. The chapter is concluded with some operational suggestions for developing written problem statements and research objectives.

RESEARCH VS. DECISION PROBLEMS

Decision and action problems (and choices about how to respond to them) often go together. We first make a decision, then act on it. Research problems and decision problems also may be closely related, especially when subject-matter and problem-solving research are involved. Disciplinary research is more likely to be directed at problems that are not directly oriented to a specific application or need for individuals or groups to make a decision or take an action. Disciplinary problems are of interest primarily within the discipline and are more likely to be oriented to the generation of knowledge without a specific decision/action purpose (although this generalization has notable exceptions, e.g., much of J. M. Keynes's disciplinary theoretical work apparently had specific policy problems at the forefront). Applied research problems more readily recognize the potential extensions and applications of research that addresses the identified problem. When research leads to a decision or action, it is potentially important to keep the research activity and the view of the research problem separated from the decision and implementation activities. The reason for this is to maintain as much neutrality as possible in the research process. Keeping them separated helps researchers maintain an objective state of mind.

To help clarify this differentiation, consider the model of problem solving presented by Glenn Johnson (1976, p. 226; see figure 6.1 below). This model illustrates the process that would include pragmatic problem-solving research. The distinction is less a problem in, for example, disciplinary research because disciplinary research is less likely to address the types of problems/issues that require decisions. Here we see the process being initiated with a problem definition, progressing to analysis, which is likely to include a conditional prescription, then someone's decision and implementation of that decision. Also represented is the use of normative and positivistic knowledge and the pragmatic interdependence between them as they impact on each step of the process. The dashed line inserted between analysis and decision in Johnson's model is this author's insertion to separate the research (information-generating) activities from the rest of the problem-solving process. Note that the feedback between the decision and action activities and the analytical activities keeps the analytical closely connected to the decision problem. This separation of research from decision and action presented here is not considered useful, or even valid, by some economists, however. Edwards (1990), for example, sees the decision process as part of the research process. He also recognizes the danger that "when a research economist becomes identified with a certain prescription, a following (or opposition) may be attracted for ideological reasons rather than for the compelling conclusions of the research process" (p. 33). Johnson (1976) may not agree with this distinction within his model. The perspective advocated here is that to mix the research and decision aspects may fragment the researcher's attention, and it may subconsciously influence the outcome of the research.

The admonishment to keep research problems separate from decision problems is not an argument that decision problems are somehow not relevant or are inferior. In fact, the decision problem, and the best answer we can get for it, is often what research users of problem-solving research are most interested in. For that reason among others, we want to keep the knowledge or information on which the decision is based as personally objective as possible. One way to foster that objectivity is to keep research activity focused on problems of knowledge, information, and understanding.

THE RESEARCH PROBLEM IDENTIFICATION

The most critical part of all research projects or plans is the identification of the problem addressed by the research. The problem is the focus of both applied and basic research activity. The reason (justification) for the research inquiry is the

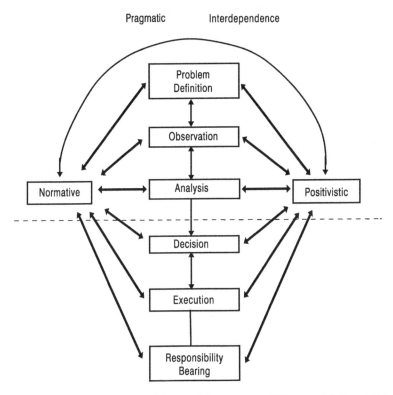

Figure 6.1.   Steps in the problem-solving process (Johnson, 1976, p. 226).

research problem. A research problem may be disciplinary, subject-matter, or problem-solving oriented, or it may be any combination of the three. In an operational mode, the problem identification and explanation affect the quality, usefulness, effectiveness, and efficiency of the research activity more than any other part of the research plan. Also, this part of the research plan is often the most difficult to formulate. This observation is based on the author's research and writing experience, working with numerous graduate and undergraduate students on their research projects, and reviewing submitted journal articles, monographs, research bulletins, and the like. Developing a clear understanding of a particular research problem and explaining it so that others readily grasp the problem is difficult. Yet it is the *reason for the research being undertaken,* and the written problem description is the only credible evidence that a clear understanding of the issue has been achieved.

One additional point before moving to more substantive issues: the individual research project addresses the problem, but the problem exists independent of the research. Researchers, and students trying to engage in research, can become confused by failing to make this distinction. When one wants to know something, that knowledge addresses some definable need if it is to be heeded by anyone, particularly when the research is not strictly disciplinary. Some of us may value knowledge for its own sake, but subject-matter and problem-solving interests have more practical needs. But even in organized disciplinary research, if the problem will not stand on its own, the proposed research is not likely to be taken seriously.

RESEARCHABLE PROBLEM VS. PROBLEMATIC SITUATION

Within the context of a specific research plan, we are most interested in the *specific* research problem that is the point of the research –that is, the researchable problem. The *researchable problem*, put another way, is the very specific problem at which the objectives of the research are directed. It is that particular lack of knowledge or information that constitutes the felt difficulty we attempt to identify, justify, and explain.

An effective way to understand the researchable problem may be to contrast it with the *general* problem, or what Williams (1984) and Andrew and Hildebrand (1982, pp. 19, 20) refer to as the problematic situation.[3] The *problematic situation* is a broad problem area or set of problems. Problematic situations are large enough in their scope that the issues embodied in them have many dimensions. Obtaining reliable knowledge and information on all dimensions often requires many years of concentrated analysis. For example, the world food (or hunger) problem is a problematic situation. Its dimensions include low incomes in developing countries, high birth rates in selected countries, national policies that discourage food production, low levels of education in some countries, restrictive trade policies, and political instability that discourages capital investment in some countries, just to name a few. Other examples of problematic situations are: environmental pollution, the balance of trade problem, poverty, affordability of health care, unemployment, and government inefficiency.

Problematic situations are not researchable within the confines of individual research efforts or projects. But specific components of those problematic situations are researchable when they are broken into distinct components. These components are confined in scope and complexity and support a set of research objectives that can be achieved within given resource constraints. This does not say that we must be satisfied with partial knowledge, but individual research efforts must be achievable in order to accomplish anything. The way we accumulate more general understanding is to put knowledge gained from specific research projects together, each having made its own unique contribution to the bigger picture, so that a more general understanding emerges. We may think of a given problematic situation as constituting a picture puzzle of interlocking pieces. A specific research project may provide information that forms one or two of its individual pieces. Through research programs—a series of planned, coordinated projects—we begin to fill in enough pieces to see the general picture, although the broader understanding may require a degree of extrapolation/interpolation and/or induction. But this is fundamentally the way that scientific knowledge grows. Perfect definitive knowledge, produced quickly, is a noble goal, but it is elusive.

THE PROBLEM STATEMENT(S)

Because the researchable problem is always linked to a more general set of problems, it is recommended that the problem statement in a research proposal be separated into two parts: the general problem (or problematic situation) and the specific problem (or the researchable problem). Note that, as was pointed out in chapter 5, the researchable problem also must have some means of being addressed. There must be data, analytical procedures, and other elements with which to address the problem. Thus, merely being specific is not sufficient. Note also that the criterion for the specific problem is not that it be "small," but rather that it be precisely specified and capable of being addressed with available resources.

The general problem statement in a research proposal is to provide the background or the setting for the researchable problem. It gives the larger picture into which the researchable problem fits—the context of the problem that the proposed research addresses. A useful rule of thumb is to introduce the problem in generalities, then progressively narrow the focus (your own attention and the attention of readers) to more specific, precise problems on which information is needed. One may choose to initiate the general problem description with a very broad problematic situation, particularly when the readers of the proposal may not already be knowledgeable on, say, an economic perspective of the broad picture. Alternatively, one may initiate the general problem description with a more confined treatment of the problem setting, particularly when the readers are likely to be already familiar with the disciplinary aspects of the problem. The problem definition, as with any written communication, needs to be tailored for the expected audience.

To illustrate, consider a typical situation encountered when a student from another country is conducting research in the United States, but the focus of the analysis is on the student's home country. In this situation the student may be defining the problem setting for his or her graduate committee, made up largely or partially of faculty members who are not familiar with the issues of the country being studied. In this type of situation, it may be advisable to provide some description of the economic, social, political, and/or institutional conditions in the country of study. This understanding may be necessary for the committee members to grasp the relevant dimensions of the problem setting. On the other hand, if the topic is on a problem, say, within the United States, and the committee is already familiar with the economic, political, and institutional frameworks, those matters may need less extensive discussion and explanation.

Consider, for example, the problem definitions in the proposals in appendices A, B, and C. The general problem background in the government agency proposal relating to effects of a type of policy structure on the part of some governments on various countries (appendix A) emphasizes governments' incentives and reasons for intervening in markets, both domestic and international, and some of the general effects of those interventions. It establishes the background against which a particular set of policy impacts will be analyzed. The dissertation proposal on pharmaceutical mergers (appendix C) explains mergers in terms of types, reasons, impacts, and trends; it addresses mergers in general as a prelude to laying the groundwork for identifying the (researchable) issues of mergers in the pharmaceutical industry. The thesis proposal on the impacts of HIV on the Russian economy (appendix B) describes the general social and economic situation in Russia—a sort of "baseline"—to frame the economic problems associated with the expanding incidence of HIV in the complex mix of forces affecting the economy.

The general problem statement sets the stage for the specific, researchable problem specification. The specific problem continues to narrow the perspective of the problem to a set of issues on which knowledge or information is needed. In this process, it is necessary, but not sufficient, to make clear what knowledge is needed. The reasons the knowledge is needed, by whom it is needed, and the potential purposes it could serve should also be clear. You must try to guide your readers' thinking in a way that they perceive the problem and need for information the same as you perceive them. In fact, if the purpose of the problem statement is to justify the objectives of the research, then the test of effectiveness of the problem definition is that readers unhesitatingly accept the research objectives. If the readers' response to the research objectives is "Of course, that's what is needed," the problem statement has effectively achieved its purpose.

The five guidelines listed below are presented as a sort of checklist to aid thinking about effective problem statements that lead to researchable problem specification, whether the problem/issue/question is disciplinary, subject-matter, or problem-solving.

A. *The researchable problem must be sufficiently specific (confined) that it can be addressed with available resources.* These resources include time, researcher expertise, data, tools (e.g., computer capacity), support personnel, and financial support. The magnitude of the researchable problem is always dependent to some degree on the resources available. What constitutes a feasible researchable problem for a researcher with a large grant, a staff of research personnel, previously accumulated databases, and abundant computer capability is obviously different from a feasible researchable problem for a graduate student without those resources. A disciplinary researchable problem must have the theoretical or methodological issue precisely identified, and the researcher must have sufficient capability to attempt the theoretical and/or methodological work. One should remember, however, that the quality of a problem specification is not dependent on how large the researchable problem is, but on how *clear* it is. As a rule, the large (broad) researchable problems are usually less clear than the confined, well-focused ones.

B. *The dimensions of the problem should be described in objective (neutral) rather than subjective (advocative) terms.* Avoid "value-judgment" language. Although the problem definition is inherently normative, discipline yourself to be precise in your description and subjugate your private values and biases. The language you choose to use may affect both the perception of how objective/neutral your research is and the reality of your neutrality. For example, if you state, "Wheat producers in the European Economic Community are overpaid," the statement implies personal subjectivity, whereas "Wheat producers in the European Economic Community are subsidized" avoids that subjective connotation. It may also help you to be more precise in your own thinking about the issue. The existence of subsidies can be substantiated (e.g., "Internal wheat prices are 2.5 times the world market price for wheat"). Whether they are overpaid is a personal judgment; how much of a subsidy constitutes "overpayment" is a matter of opinion.

Avoiding subjective language in defining the problem may prevent the discussion of the problem focus from drifting into areas that have no answers. Also, avoiding unnecessary or unneeded descriptors, particularly those that express an opinion, is also advisable. For example, "Unfortunately, 20% of small businesses go bankrupt each year" is a statement for which the addition of *unfortunately* adds no substantive understanding; it connotes an emotional appeal.

C. *The problem must be described sufficiently that other people (the audience) comprehend it.* This applies to both the general and the specific problem statements.

D. *The researcher's perception of the problem may be derived from an intuitive level, but it should be developed to a logical level.* It is only in this manner that the researcher and others may really understand the problem. If the description of the problem is not developed logically to a point that others grasp it, then the problem is not well defined. Emotional statements fail to offer evidence.

E. *The inquiry may be initiated with a research objective—something that you want to determine—but the problem explanation should be developed to provide the justification for the objective.* If you cannot explain the problem that is the reason (justification) for the objective(s) of the research, then either you do not clearly understand the problem yourself, or there is no problem.

USE OF DATA

In the processes of understanding and describing problems, several important components may be involved. We have noted that our perceptions, conditioned by personal experiences, are likely to be involved. Theory also plays an important role in understanding the problem, as will be discussed in chapter 8. Yet another component of understanding the problem, be it a problematic situation or a researchable problem, is data (Edwards, 1978, 1990).

Data, and the process of collecting them, can affect how one perceives a problem and one's awareness of a problem. Data often are instrumental in revealing problems that might have otherwise gone undetected. Consequently, it is usually advisable to seek out and examine relevant data in the process of developing the problem statements. If your research topic includes employment considerations, go to the library and locate appropriate data series on employment statistics. If your topic includes trade in automobiles, locate data series on automobile production, exports, and imports, perhaps in different countries. You may be surprised at what the data reveal, and how the selected inclusion of some of those data in your problem definition can foster the readers' comprehension of the problem. Edwards (1978, p. 31) may best state the case:

> In the descriptive process, people arrange the available data to tell a story and to illustrate a problem. The person who understands the ultimate social problem must sift through, classify, aggregate, and rearrange the data and reduce large volumes of miscellaneous facts to manageable information. This interaction between unorganized data and social problems results in meaningful descriptions. It carries us beyond vaguely felt needs toward precise conceptualizations of a problem and provides for measures of its nature and extent.

Descriptive research, or the process of synthesizing information, often includes the interpretative use of data that may be included in problem definitions, as discussed in chapter 2. Put another way, defining research problems often includes conducting descriptive research, that is, synthesis. Note, for example, how the proposal in appendix C uses data to show changes in market structure in the pharmaceutical industry.

OBJECTIVES

The objectives specified in the research proposal address the specific, researchable problem. They specify what the research project proposes to accomplish (do, achieve, determine, estimate, measure, evaluate, etc.). Objectives are usually best specified in the form of (1) a general objective; and (2) a set of specific objectives (refer to the objective statements in appendices A–C). Although the statement of objectives is often the shortest part of any research proposal, it remains the centerpiece. The other parts of the proposal support and give credence to the objectives.

The *general objective* is the central objective of the research study. It states the main purpose of the research. It also should flow directly from the statement of the researchable problem. For example, if the specific problem is that little is known about the response of toxic waste emissions into the air from government fines for toxic waste emissions specified by particular legislation, then an appropriate general objective might be to determine how the legislated schedule of fines has affected, or would affect, the concentration of toxic waste emissions in the northeastern United States. The more succinctly and precisely the general objective can be stated, the better its function is served. A useful guideline is that general objectives should be stated in one sentence.

The *specific objectives* constitute a set, perhaps a list, of subobjectives that contribute to the general objective. The achievement of each specific objective contributes something toward achieving the general objective, and the primary objective is achieved in total or in part by achieving the set of specific objectives. Sometimes additional analysis that involves combining results from specific objectives is needed as well.

Continuing with the above example of a general objective, appropriate specific objectives might be to: (1) identify the legislated schedule of fines specified for toxic waste polluters; (2) estimate the responsiveness of toxic waste producers in the northeastern United States to various levels of fines (and possibly to the standards of enforcing the fines); and (3) determine the expected response of toxic waste emitted to levels of fines other than those specified legislatively. Note that the researcher must have already done enough background work to expect that the objectives are achievable (the problem is actually researchable). This may include conceptual and preliminary modeling work, exploration of data sources, and knowledge of prior studies related to the topic.

DEVELOPING PROBLEM AND OBJECTIVE STATEMENTS

As is usually the case in creative matters of this type, there are no specific, unassailable rules on how to develop the statement of the problem and the objectives in a research proposal. Furthermore, what works for you in one set of circumstances may not work the same again when the situation varies. The suggestions that follow are derived from having struggled with the process and having worked with students who were struggling to develop clear, effective, efficient statements of problems and objectives. You may start with either a broad subject-matter area in which you are interested in conducting research or a specific researchable problem or a general objective, as noted above. Suggestions are segregated on the basis of the way in which you initiate the research inquiry.

If you begin with a broad area in which you want to conduct your research, remember that the task is to get that research interest narrowed to a researchable problem. Start by segmenting the problem area into its logical components, then continue to segment those into smaller parts (issues, questions, etc.). Unless you have a natural affinity for the process or are very lucky, you will likely need to work through multiple iterations of the process. Discuss the questions, needs, issues with peers and advisors; explain what you already understand and try to explain where you are stuck. Sometimes the act of verbalizing your thoughts gives more clarity to them. Explore all aspects of the problem, and at all levels of specificity. Remember, however, that you probably will not want to include everything you have thought of in your written statements. Your written statements need to follow a direct, linear train of thought. Getting sidetracked will only confuse your readers and evaluators. Follow a progression in your explanation—sequentially narrow the reader's focus to a progressively specific problem.

If you start with the opposite perspective—you already have an idea of your researchable problem (or general objective)—then you must develop and explain the more general dimensions of it so that readers and evaluators can also understand the context and setting of the researchable problem. The task is to "backtrack" through the logical sequence leading to the justification of the researchable problem. Starting from the statement of the researchable problem, ask yourself why the researchable problem exists (why the knowledge is needed), to whom it is a problem (who needs the knowledge or information), and how it became a problem. Proceed by asking these questions at the next level of generalization, and proceed to the level of generalization needed for a starting point. Remember that you are searching for a point at which to initiate your description of the general problem or problem situation. Your focus can become too general—too elementary for the audience. Once you can identify a starting point for the general problem statement, then reverse the process and fill in the logical sequence in the same manner as above.

Once the problem has been defined, framed, and justified, and the general objective therefore identified, the specific objectives become a matter of identifying an appropriate set of subobjectives that leads to the general objective. Ask yourself what component parts must be determined or understood in order to achieve the knowledge required for the general objective. For example, if determining an economic optimum for a procedure, process, or policy is a general objective, then determining the costs and benefits associated with the procedure, process, or policy are likely to be specific objectives. You may find that the identification of specific objectives is rarely mentally or conceptually demanding, but it requires attention to details.

A note of caution on specifying specific objectives: avoid confusing objectives with methods. The specific objectives address *what* is needed to achieve the general objective, not *how* the general objective (or specific objectives) will be achieved. Students often want to state a specific objective as "To construct a model of . . . " when constructing a model is a means of achieving an objective (a method or procedure), not the objective itself, especially in subject-matter and problem-solving research. Granted, the two may go together, but what perspective has the researcher lost by confounding means with ends? Developing new methods *may* constitute an objective, but that situation identifies a purely disciplinary activity. Thus, the separation between objectives and procedures may, on occasion, be violated, but the instances when a model is a technically correct objective are limited. Keeping objectives separated from the methods and procedures remains the best *general* approach.

Another note on specifying and writing problems and objectives: do not expect this exercise to be a straightforward, once-through endeavor. Unless you are experienced and already have done extensive research in the topic area of the proposed research, expect to rotate back to the problem statement and objectives repeatedly as you gain more background and perspective: for example, as you proceed through more literature search, conceptual thinking, and data. Consider that the problem statement presents a perspective, interpretation, or point of view about a problem. It is therefore not unusual that you would alter your perspective, and your explanation of it, as you gain more understanding of matters that are necessarily related to the problem. Do not be surprised if your understanding and description of the problem are altered substantially from the time you initiate your research project until you complete it.

Most researchers find that developing a logical, clearly written statement of the problem and objectives is the most difficult and demanding aspect of proposal writing, if not of the entire research process. It requires clearer thinking and more communication skills than perhaps any other part of the research proposal. Yet proposals may succeed or fail most often based on their problem specification than on any other single component. Few reviewers feel comfortable in judging the substantive procedural aspects of proposed research unless or until they understand the problem. The technical aspects of your proposed research are unlikely to receive serious consideration without a clear problem description. A good problem statement will not, however, rectify incomplete or poorly thought-out procedures.

It is difficult to overemphasize the importance of peer and advisor review and input in the process of developing problem statements and statements of objectives. Interacting with peers and advisors, both orally and through written drafts, is not only a device for increasing the efficiency of our communications, it is necessary for many of us to clarify our thinking. You may find it useful at this point to review the suggestions on writing in chapter 5.

SUMMARY

This chapter considers the definition of research problems and their associated objectives from the perspectives of both formulating them and writing them for a research proposal. A distinction is made between research and decision problems, and the reasons for differentiating between them in the research process are discussed. Research problems are often defined by the need to solve decision problems. Conversely, the solution of problem-solving and subject-matter research problems may, and hopefully often do, provide the knowledge or information with which decision problems are addressed.

The research problem is the focal point of research activity. We can visualize it as having two components: the *general* problem, or problematic situation, and the *specific* problem, or the researchable problem. It is the researchable problem delineation that is most important for focusing the individual research project or activity. However, specification of the more general problematic situation framework is necessary to place the researchable problem in context.

Objectives are also presented as needing to be specified in general and specific terms. General objectives identify the overall intent of the research project, while specific objectives identify the subparts that will contribute to achievement of the general objective. Problems with confusing objectives with methods and procedures are discussed.

Guidelines for writing the problem statements are provided. These include considerations of specificity (confinement of the researchable problem), language use, and use of empirical data to demonstrate problems. Guidelines on writing objectives are also provided.

SUGGESTED EXERCISES

1. Obtain a research proposal (e.g., from a student, faculty member, or research organization). Read the problem and objective statements. Can you identify the problematic situation, the researchable problem, the general objective, and the specific objectives? Explain.

2. Select two journal articles from recent issues of economics journals, preferably different journals. Explain the extent to which each addresses research objectives and decision objectives as they are explained in this chapter. Can you identify the problematic situation and the researchable problem in each? Can you identify subjective language in the problem statements (or in any other part of the articles)?

NOTES

1. A *problem*, as viewed in this context, may also be labeled *issue* or *question*. Readers may substitute these labels as desired.

2. Back (no date) noted that it is curious that Northrop omitted problems of method dealing with how to generate reliable knowledge.

3. Breimyer (1991, p. 247) attributes the original delineation of the problematic situation as separate from the research problem to Salter (1942).

7

The Literature Review

Before we can advance the state of knowledge, we must first know the status of the current state of knowledge. Whatever the research topic, others have conducted research that is related to that topic. The effective researcher will use the knowledge and insight of others who have done research on the same topic or in ancillary areas. Examination of the development of economic thought, and the evolution of science in general, shows that even the notable "breakthroughs" in science and research have drawn from prior work and used elements of prior research done by others. For example, Alfred Marshall, who formalized in 1890 much of what is still recognized today as modern microeconomics, drew heavily on the research and writings of the classicists, utopians, and marginalists who preceded him. Marshall was a synthesizer to a large degree; he took elements of economic thinking from the preceding century and formulated something unique from it. The history of science and philosophy abounds with these examples. Rudimentary ideas underlying some concepts developed and formalized in the 1930s by John Maynard Keynes can be traced back to Thomas Malthus more than a century earlier.

Thus, the task in the literature review is to learn as much as you can from the efforts and work of others. This activity, however, cannot be haphazard if it is to be efficient. Like most aspects of research, it needs thought and planning. This chapter explains the review of literature and its objectives, gives guidelines on how to proceed with it, and provides suggestions on writing a formal literature review. The chapter is organized as follows: a discussion of the purposes and functions of the literature is presented, then guidelines on how to proceed with the search are given. Writing guidelines are offered, followed by a final section on referencing. Occasional reference to the literature review sections of the proposals in appendices A–C will be useful throughout this chapter.

PURPOSES OF THE LITERATURE REVIEW

The central purpose of the literature review is to provide the researcher and the reader with an understanding of the body of literature as it relates to the current or proposed research. The purpose includes an understanding of the strengths and weaknesses of the prior research. Your research problem is the focus of your literature review, but the literature you review may be related to your research project in several different ways—through the problem, the objectives, the conceptual framework, and/or the methods and procedures. Prior research that addressed a similar problem or had similar objectives is related to your research because knowing how others approached their problem and objectives is relevant to your investigation. Likewise, knowing the approaches of prior researchers may suggest feasible or infeasible approaches or procedures for your work. The conceptual reasoning of others regarding a similar or related problem may provide the foundation for the conceptual reasoning about your own problem. Alternatively, it may provide a basis for new conceptual developments.

It is important to remember that other studies may be related via both similarities and differences; the differences may be particularly relevant with respect to methods and procedures. For example, someone else may have followed an approach substantially different from the approach you propose for a similar set of objectives. Knowing their results, and the success or lack of it from their approach, is still important for your approach; it provides the basis for comparison of the effectiveness of the alternative approaches and of other research results.

In addition to the central, or general, purpose of the literature review identified above, we can list several other, more specific, purposes, functions, or benefits. This list draws from Leedy (1985, p. 69) and Ary, Jacobs, and Razavieh (1979, pp. 57, 58). The literature review

A. *Prevents unwarranted or excess duplication of what has already been done.* Some duplication or confirmation of research findings is usually necessary, but reproducing a study that has already been well established is likely to cost more than its potential gain.

B. *Helps to identify the frontiers of the field and helps to identify how, where, and in what manner the proposed research might add to the general body of knowledge.* It helps the researcher and the proposal reviewer to understand how the research would contribute to the more general understanding of the problem and the field of study. It also helps researchers to interpret the significance of research results after the fact.

C. *Provides ideas and direction for*

1. how to handle problems encountered.
2. techniques.
3. sources of data.
4. approaches that may not have occurred to the researcher.

D. *Helps develop insight on the design of one's own study by showing what approaches were and were not successful in prior analyses.* Additionally, it may show which approaches have and have not been attempted.

E. *May reveal conceptual insights into the problem and/or provide the basis for hypotheses in the researcher's own study.*

The role of the literature review is pivotal in any piece of research, irrespective of its nature. The review is as important in problem-solving and subject-matter research as it is in disciplinary research. One cannot advance the state of knowledge/understanding without knowing what the current state is. As noted in chapter 5, you are not likely to finalize a problem definition and objectives until your literature review is completed.

A formal literature review may not be required in all research proposals. For example, sometimes problem-solving research proposals aimed at industry groups may not require a formal literature review, but rather a brief recap of the current state of information related to the problem with some references—much like the hurried summary of related studies that is common in professional journal articles. An example of this approach can be seen in appendix A, where much of the literature is cited in the problem definition. Whether a formal written review is required or not, you should never undertake any research activity without knowledge of the literature related to your research. To do otherwise imposes risk of unneeded duplication, repeating avoidable mistakes, and ineffective or inefficient research. The literature review is not merely something that must be done before engaging in the substance of the research project. It is a necessary, integral part of the process of planning and designing effective, relevant research.

THE LITERATURE SEARCH PROCESS

The first point regarding the search for related research literature is that *all literature is not eligible for inclusion.* This limitation is not to suggest that reading other literature is not important. In fact, literature such as trade journals and business and commercial news is essential for understanding issues in the world about us. However, only "scientific" literature—literature that has been through a peer-refereed or review process—should be included in the literature review portion of a formal research proposal. This includes all types of academic and professional journals and most formal research reports produced by the staff of or under the sponsorship of government agencies, university-affiliated bulletins, reports, and monographs. It also includes reports by research foundations and organizations, international organizations, and similar outlets. The specific types and outlets are much too numerous to list in specific terms.

The fact that a publication is refereed or reviewed literature does not constitute evidence that it is infallible, or even correct. It does, however, indicate that there have been some independent checks on accuracy, validity, and correctness. This is not the case with published material that has not been reviewed for analytical substance by persons with qualifications to evaluate such matters. Examples include all types of "popular" publications, such as newspapers, news magazines, and publications by lobbying organizations or groups. Even sources that tend to be respected for journalistic accuracy (e.g., *The Wall Street Journal, The Economist,* and *Barrons*) are not appropriate sources for the literature review. This does not say that such material might not be used for background perspective in, say, the problem description, but it should not be included in the literature review. The literature review is intended to provide an overview and summary of prior reliable knowledge, and the reliability of the popular literature lacks a system of checks.

A necessity for conducting research, including (but not limited to) the literature review, is access to a research library. To cover the range of specialized literature necessary for research, the library must have most of the specialized search facilities and capacity available only in research libraries. Your library does not necessarily have to have every publication you may ever need on the shelf, but it must have the facilities to locate the references to anything/everything you need and be able to obtain the material if it is not in the library in some form. No individual library has all sources within its walls, but research libraries can access information through other libraries. More of the interlibrary accessibility is being conducted through electronic communication, and it is not difficult to foresee that documents will, in the future, be stored in computer memory and accessed from practically any remote location through telecommunications networks. The library of the future may contain relatively little of its information in paper or hard-copy form.

Exhaustive literature reviews and searches become increasingly difficult as the total volume of literature expands, and it is expanding exponentially. Irrespective of whether the knowledge base is expanding that rapidly or we are merely publishing too much material of minor importance, the increased volume and proliferation of publications makes thorough literature reviews very difficult, but they are nonetheless necessary. Otherwise, we collectively engage in unneeded or inefficient duplication. This problem makes the use of literature search aids all the more important.

Search Aids

Standard search aids include indexes, abstracts, and bibliographies. There are many of these search tools available in any research library, some of them computerized.[1] Computerized literature search capability provides the means to save considerable time in identifying potential sources and is expanding rapidly. It is, however, too limited in the total volume of material indexed at the present time to be used as a sole means of search. Until such time when all published material is included in computer indexes, both computer and hard-copy searches will be required. You also should not overlook search aids that cover literature stored specifically on microfilm or microfiche.

An additional caution: *It is not advisable to rely solely on the search aids to locate all relevant literature.* Most of the indexing and abstracting services lag behind the current literature by several months to as much as two years. If you rely solely on the indexes, abstracts, and the like, you may overlook more recent related literature. One way to manage this problem is to check current issues of major journals and other types of outlets, relying on the organized search aids to identify the most likely outlets for literature related to your topic. An excellent search aid that may be overlooked is the dissertation abstracts available through most research libraries. Dissertations and theses often contain details and explanations of analyses that are omitted in journal articles and other condensed publication outlets that are derived from the theses and dissertations.

Key Words

In initiating the search for the related literature, selecting a set of key words is necessary, and the time spent doing this well can improve your efficiency. All literature search through the standard search aids uses a system of key words to locate published sources. If the set of key words you select is too confined, you may overlook or fail to locate relevant literature. If the set is too broad, you may spend time locating and reviewing sources that have no discernable bearing on your study. The best advice is to be cautious in the initial stages; that is, be broader in your list of key words, then narrow it to a more confined list as you progress and develop a "feel" for the key words that are getting you to the literature that is appropriate. This approach will cause you to search out some unnecessary material early in the search, but the risk of having to repeat part of the search, which is even more time consuming, is minimized.

Guidelines on how to define key words are difficult to formulate; it is similar to the process of selecting the key words for something you have written. Try to focus on words or short phrases that describe the topic you are studying. Remember that you may need to cover literature related to your study through the problem, objectives, conceptual framework, and/or methods and procedures. Consequently, you may need to review prior research that used certain analytical techniques as well as research that has addressed certain types of problems or issues. For example, if your study proposes to analyze the optimum use of the water resources in a specific river basin and proposes to use dynamic programming techniques in the process, then appropriate key words probably should include dynamic programming (technique-oriented literature) as well as water allocation, water use efficiency, river basins (subject-matter–oriented), and the specific river basin (problem-solving–oriented). You need to be aware of prior research on the specific river basin, but you also need to know about research on other river basins because some of that information may be transferable.

As you locate titles and abstracts, you must make an initial determination regarding the potential relevance to your study. This task is easier when an abstract is available. When only a title is available, you must decide whether to locate the publication and examine it without any further information; this demonstrates the importance of titles that are descriptive of the content of the research papers. The suggested decision rule is: when in doubt, be conservative; in research, it is better to select thoroughness over convenience.

Reading

It is often useful to start your reading with the most recent studies. Two things are accomplished: (1) you focus more quickly on the current state of knowledge and understanding; and (2) the recent research often includes references to relevant earlier research. As you examine each source, first read the abstract and/or summary of the article or report. This permits you to determine the relationship, or lack of it, to your study and whether you need to review it for inclusion in your literature review. This practice can save time. The suggested guideline is, as usual, when in doubt, err on the safe side. It is usually more efficient to have read something that you may eventually elect to leave out of the formal literature review than to have to search out the paper again, review it, and insert its contents into the literature review. You also want to avoid the prospect of having omitted a potentially important source from the literature only to have someone call it to your attention later, or at the conclusion of the research project.

As you read, keep in mind that your central purpose is to identify and describe the study's relevance to your study. Your reading broadens your perspective, and may even cause you to rethink and revise your problem and objectives, but you must keep the central purpose in view—how the prior literature relates to your proposed research.

Notes

Be certain that you have a complete citation on each source. Students often waste time because they neglect to note minor items needed for citations or references—items like year of publication, page numbers, and city where a book publisher is located. (More will be covered on referencing in a later section.) Also, make sure that the correct reference is retained with the notes on each source you review.

Keep written notes; do not rely on memory. Even if you think you will remember, material from many readings will begin to confuse you on which idea or thought is associated with which study or author. Be thorough and systematic in taking notes. Note the problem, objectives, methods, findings, and conclusions from each study reviewed. Pay particular attention to how each source is relevant to your study. Note any questions, perceived shortcomings, or problems not addressed of the studies reviewed. These things will be helpful later both in collectively organizing your thoughts about the literature and in writing your review of literature.

Many authors advise the use of note cards in organizing the literature reviewed. While this procedure is certainly reliable, you may find that using standard-sized paper accomplishes the same for your purposes. You can adopt whatever specific tools with which you are comfortable, as long as you are systematic enough about it to facilitate an organized, useful review.

WRITING THE LITERATURE REVIEW

The initial task for most researchers in writing the literature review is to get the right attitude. The process is not a mere technicality or an unnecessary addendum to the proposal. It serves the vital, necessary purpose of describing and critiquing or analyzing the prior work that is related to your proposed study. It gives perspective on the proposed research, provides support for the problem description and objectives, and sets the tone for the conceptual considerations and methods and procedures that follow.

To achieve its function, the literature review, like any other part of the proposal, needs organization. It is not merely a series of unconnected summaries of studies. It is ideally a synthesis of the previous related research. The manner in which it is organized may vary, but it requires organization to communicate effectively. You should develop an outline of the review before starting to write. Initiating the literature review with an introduction that sets the tone and prepares the reader for what is to follow is recommended. A short ending summary section of the literature review that pulls all the material together and leaves the reader with a clear delineation of what the author concludes about the total body of related literature may also be a productive addendum.

Use of subheadings in the literature review often helps to organize the presentation. These are usually arranged under some subject-matter headings—logical groupings of studies with a similar focus, such as that illustrated in appendix C. These groupings will vary with the purposes at hand, and sometimes a single piece of literature may need to be discussed under more than one subheading. Within a grouping, the material can be organized in chronological order, or it may be best presented by constructing the story along other lines. Whichever way it is presented, the relation to your study should be clear to the reader. Recall that differences with, say, the approach proposed for your study may be as important as the similarities. Both you and your readers need a clear understanding of the significance of each source covered to the study under consideration.

You should summarize—not repeat—then analyze, compare, and contrast the literature reviewed. The purpose is not to explain every detail of the previous research, but to ensure that your readers grasp the main aspects of the prior research as it relates to your proposed research. Providing analysis and critique of the strengths, shortcomings, and contributions of prior studies, both individually and collectively, is important in the literature review. Your description can be too short, but it can also be too long. The recommendation is, again, when in doubt, write more rather than less. The reason for this recommendation is that in editing, it is easier to condense a description than to augment an incomplete one. While quotations may be useful in describing the prior work, avoid long quotations in the literature review. Also, be reluctant to reproduce any graphs or tables from prior literature; do so only when the general conclusion can be obtained only with those devices. Restating the details of prior research is not the mission of the literature review. If the reader desires details of any of the literature reviewed, you provide the means for him or her to obtain it with the references to the literature you provide.

When writing the literature review, do not overlook the economic foundation literature related to your research. That is, you may need to provide an overview of the progression of developments, near-term or long-term, in conceptual thinking, analytical procedures, and empirical evidence, in order to place your research in its relevant perspective. For example, review of literature for a project that involves hedonic price analysis may need to indicate and briefly summarize the theoretical foundations provided by Lancaster (1966) and Rosen (1974), and possibly others such as Ladd and Suvannunt (1976). However, one might also need to trace the earlier empirical works going back to Griliches (1961), Waugh (1928), or even Taylor (1916). It is not suggested that one always needs such a perspective on foundation literature. However, its omission can, in some cases, limit an otherwise instructive literature review.

A sometimes perplexing problem in writing the literature review, as well as writing in general, is when to summarize the content of a published source and when to merely refer to it. Specific guidelines are difficult because the decision depends in part on the intended audience. More summary is required in a graduate thesis or dissertation proposal than in, say, a proposal for an industry group or perhaps some proposals for government agencies. When in doubt, and the situation makes it possible, ask.

A related note of caution: *Do not reference a source of an idea without having actually read the source.* It is common practice to take "shortcuts" by relying on someone else's interpretation and summary of an original source. However, the problem with that approach is illustrated by the game in which a person relates a story to another person, then that person relates it to a third person, and so on. After seven or eight exchanges the story is related back to the original person. The final story related rarely bears much resemblance to the initial story. Another person's interpretation and summary may differ from, and be inferior to, your interpretation. If one reads the original source, these types of errors may be minimized.

REFERENCING

Although referencing is used throughout research proposals and research reports, it is discussed in conjunction with the literature review because it is used most frequently there. The style and procedures of referencing should not vary within a document.

We reference other works for several reasons. One is to provide supporting evidence (or evidence to the contrary) for the propositions, results, and views that we ourselves offer. Another reason is to assign responsibility for an idea, position, concept, or result. To provide readers with added information and detail on matters discussed or summarized in our own writing is also an important reason. Still another reason is to give credit for the thoughts, ideas, efforts, and contributions of others. This is an ethical issue of major importance in professional and academic circles that warrants some elaboration.

Failure to attribute an idea or research result to its originator constitutes *plagiarism*. It is the presentation of another's words or ideas as your own (Babbie, 1998). The way to avoid this most serious professional sin is to provide adequate, accurate referencing. Students sometimes develop the attitude that having to attribute ideas or findings to others somehow detracts from the significance or originality of their own efforts. WRONG! It more likely shows an ability to effectively collect, utilize, arrange, and synthesize the information developed by others. This, in turn, suggests an increased probability that the efforts to expand the body of information or knowledge will be successful. Effectively using the works of others has been an ability of many great thinkers and researchers throughout history. None of us achieves much of significance without benefit of the efforts of our predecessors and our peers. It is difficult to reference other works too heavily. When in doubt, it is best to recognize others' contributions.

The Committee on the Conduct of Science, National Academy of Sciences, made a statement concerning referencing and plagiarism that is worth noting:

> Special care must be taken when dealing with unpublished materials belonging to others, especially with grant applications and papers seen or heard prior to publication or public discharge. Such privileged material must not be exploited or disclosed to others who might exploit it. Scrupulous honesty is essential in such matters. Even though plagiarism does not introduce spurious findings into science, outright pilfering of another's work draws harsh responses. If plagiarism is established, all of one's work will appear contaminated. (Ayala et al., 1989, p. 18)

To avoid plagiarizing, give credit whenever you use:

1. Another person's idea, opinion, or theory.
2. Any facts, statistics, graphs, drawings—any pieces of information that are not common knowledge.
3. Quotations of another's spoken or written words.
4. Paraphrase of another's words. (Indiana University, 1998)

Keep in mind that the principal intent of referencing, regardless of the specific style used, is to provide sufficient information for the reader to locate the specific source. The style used in referencing may vary with the type of outlet, the specifications of agencies or publication outlets, or personal preference. One style is to list references in *footnotes*. Footnotes are typically identified with a numerical superscript, using successive numbers throughout the paper (or starting over with the number 1 on each page), with the citation at the bottom of the page.

Footnotes are not the preferred style for many professional types of outlets, possibly because of the difficulty of organizing and placing them on pages. However, the widespread use of word processing software packages on today's microcomputers minimizes these problems. Readers may consult various reference sources on technical writing for more instructions on footnoting and other reference styles.[2]

Another style, which is really a variation on footnotes, uses *endnotes*. Endnotes are used in the same manner as footnotes, and in the same form, but the notes are placed at the end of the paper rather than at the bottom of the page where the citation is given. Endnotes serve the same function, but have the advantage of avoiding placement problems. Their disadvantage is that the reader must turn to the end of the paper to locate the source. Journal editors often specify endnotes instead of footnotes in submitted manuscripts.

The most widely used style for referencing in professional publications and papers is some version of *parenthetical reference*. With this approach the reference is provided with the author's last name, plus year of publication and page of the publication when appropriate. The full citation is listed at the end of the paper under "References" or "Bibliography." The difference between a reference list and a bibliography is that a bibliography may contain citations that are not referenced specifically in the paper. The citations might be given as (Andrew and Hildebrand), (Andrew and Hildebrand, 1982), or (Andrew and Hildebrand, 1982, p. xx), for example. The year of the publication may be considered optional unless the authors have more than one citation within the manuscript. If the author has more than one citation in the same year, the years have an added identifier in the citation and the reference list (e.g., 1985a, 1985b). The page number is added when there is a need to be specific on the location of the idea, thought, or quotation. The parenthetical reference may omit the author's name from the parentheses when the name is used in the sentence. Examples: "Andrew and Hildebrand noted that . . . ; Andrew and Hildebrand (1982) noted that . . . ; Andrew and Hildebrand (1982, p. xx) noted that . . . ". The system of parenthetical citations is often preferred for referencing style because it is easiest to use. It is also specified by most economics and by many other professional journals.

A less widely used variation of the parenthetical reference is to use numbers in place of the author's last name, with the corresponding number preceding the source in the list of references. For example, (4) or (4, p. xx) would be placed in the appropriate spot in the sentence. Few publications use this style because it poses a logistical problem of having to revise numbers, and keep the numbers correct, as references are added or deleted.

There are some variations on the format for references in the bibliography or list of references. The following provide some illustrations.

For *journal articles*:

McCloskey, Donald N. "The Rhetoric of Economics." *Journal of Economic Literature.* XXI (June 1983): 481–517.

In the above example, the information denotes vol. XXI, year 1983, pages 481–517; the number within the volume and/or the month may or may not be included in the reference. An acceptable alternative for journal articles is:

McCloskey, Donald N. 1983. "The Rhetoric of Economics." *Journal of Economic Literature.* XXI: 481–517.

For *books*:

Blaug, Mark. *The Methodology of Economics, or How Economists Explain.* New York: Cambridge University Press, 1980.

For *chapters in books*:

Johnson, Glenn L. "Holistic Modeling of Multidisciplinary, Subject Matter, and Problem Domains." Chapter 5, *Systems Economics*, Karl A. Fox and Don G. Miles, editors. Ames, IA: Iowa State University Press, 1987.

For *technical bulletins, monographs*:

Ghebremedhin, Tesfa, and Luther Tweeten. "A Guide to Scientific Writing and Research Methodology." Agriculture Policy Analysis Project, Oklahoma State University, Report B-26, Feb. 1, 1988.

Some types of references can be confusing and/or require some ingenuity or creativity. There are occasions when you may need to reference sources that are not published. Some examples follow.

For *unpublished manuscripts*:

Williams, Willard F. Unpublished class notes for a course in research methodology in economics at Texas Tech University, 1984.

For *personal interviews*:

Doe, John Q., Vice President for Operations, XYZ Corporation. Unpublished data from manufacturing operations of XYZ Corp. obtained in an interview, November 28, 1991. Information used with permission of the firm.

Data references are similar to other references. When the data are published, the source often does not have an individual person or persons as author. The rule for that case is that the organization be listed as the author. For example:

United States Department of Commerce. *Survey of Current Business.* Bureau of Economic Analysis, Wash., DC, Vols. 40–69 (1961–1989).

Or

International Cotton Advisory Committee. *Cotton World Statistics.* Wash., DC, Vol. 44 (1990).

If data are unpublished, the citation might be similar to the personal interview citation above.

For computer software, the citation may refer to the user's manual for the software. To reference use of the Statistical Analysis System, for example, the citation could be:

SAS Institute Inc. *SAS/ETS User's Guide.* Cary, NC, 1984.

Econometric, simulation, optimization, or other types of models may be more problematic. If the model is documented (i.e., its structure is specified either in mathematical or computer-language terms, in written form), the written document is referenced. If it is not documented, one can only attest to its existence. This type of "nonreference" presents the profession with some procedural dilemmas.[3]

There are numerous acceptable forms and formats for referencing. There are two guidelines that will serve you well. First, *check to determine if there is a specified style for the particular outlet* (e.g., funding agency, journal). The style should be identified and selected before the formal writing process begins. Second, *for whatever style you use, be consistent in its use throughout the document you are writing.*

The expanded use of sources available on the Internet poses new situations with referencing that are not yet fully resolved, even among professional librarians and writers. A growing number of refereed journals are available on Web sites and some *only* in electronic form, and the citation form/format has not been totally resolved. However, what seems to be acceptable is the inclusion of the full Web site address in the citation, either in addition to the print citation or in place of the print citation, as appropriate. For example, sources for discussions of citation issues for electronic sources include Page (no date), Walker (no date), and Li and Crane (1996) in the list of references at the end of this book.

SUMMARY

This chapter concentrates on purposes of the literature review in a research proposal, suggestions on how to approach the search for relevant literature, suggestions on organizing the literature review, and referencing the literature used in it. The most basic purpose of the review is to document and summarize the prior scientific work that is related to the proposed research. In achieving that, more specific purposes may include avoiding research duplication, identifying frontiers, directing procedural aspects, and identifying prior successful approaches. The literature review makes contributions to each of the other components of the proposal—problem and objective identification, conceptual framework, and methods and procedures.

The literature search process considers only sources that have a formal, rigorous review process associated with them. Popular or journalistic sources, while potentially useful in problem descriptions and for general understanding of perspectives and issues, are not used in the proposal literature review. Access to a research library is necessary and use of all types of search aids, computerized and otherwise, is advised. As the search proceeds, invariably with the use of key words, documentation of sources and taking appropriate notes help to organize the process.

The written literature review, like other parts of the proposal, needs planning and organization. Its organization helps to educate both the writer and readers. Organization may vary with the topic and intent of the proposal and the preferences of the researcher. Nevertheless, some writing suggestions are offered.

Referencing is used throughout the research proposal but is included in this chapter because of its extensive use in the literature review. The various reasons for referencing are identified. They include providing supporting evidence, assigning responsibility and credit, and providing information. The ethical issue of plagiarism is raised. Referencing styles are also discussed, with examples of common styles in economics provided. Numerous styles are acceptable in various outlets, but a set rule is that the style does not vary within an individual document.

RECOMMENDED READING

Leedy. "The Review of the Related Literature." Chapter 4, *Practical Research Planning and Design*, pp. 68–78.

SUGGESTED EXERCISES

1. Go to your research library and select two theses or dissertations. For each, read its problem statement and objectives, then read and critique the literature review. In your critique, pay specific attention to (a) the aspects of the research that were covered (problem, objectives, concepts, methods and procedures); and (b) the understanding of the review of literature that the writer conveys to the reader (e.g., does the writer provide enough information and understanding so that you, the reader, can proceed with confidence that the writer has a command of the research related to his or her research?).

2. Students taking English composition courses, typically as freshmen, are assigned "research papers" that consist of locating and summarizing material in a library. Provide a brief explanation of the relationship of that type of "research paper" and a literature review in economic research. Orient the explanation so that it is understandable to the typical English composition instructor.

NOTES

1.   There are also services that index or list research projects in progress. For example, the U.S. Department of Agriculture (USDA) maintains a computerized search service, Cooperative Research Information System (CRIS), that lists all current research projects within USDA and active research connected with USDA being done at universities, with contractors, and the like.

2.   It should be noted that footnotes may be used for more than just referencing. They are also used to provide explanation or elaboration on points that might interrupt the general flow of thought of the writing. They are also used to provide short mathematical proofs, for example, rather than including them in the text. Most outlets recommend judicious use of footnotes, however.

3.   When large and complex economic models are developed at substantial cost, developers of such models often want to protect their proprietary or financial interests in them. If they were fully documented, others would use them almost as easily as the developers. The dilemma arises in establishing the validity and scientific credibility of such models. Having model results without knowledge of how they were produced leaves one without the means of (1) deciding how reliable the results may be; and (2) reproducing the results. It is as if those who design, operate, or use the models say, "I cannot show you how the results came about, but trust me; they are reliable." These models, when used without documentation of the specific structure, are sometimes referred to as "black boxes."

8

The Conceptual Framework

Economics is a science of thinking in terms of models joined to the art of choosing models which are relevant to the contemporary world. Good economists are scarce because the gift of using "vigilant observation" to choose good models, although it does not require a highly specialized intellectual technique, appears to be a very rare one.

John Maynard Keynes

As noted previously, a formal conceptual framework is not required in all research proposals. Whether required or not, formal or not, no economic research endeavor should be undertaken without benefit of the researcher(s) having developed a conceptual framework for the project. The objective of this chapter is to define, explain, and provide guidelines for developing the conceptual framework within a research proposal or project. In this chapter it is assumed that a formal, written conceptual framework is a part of the written research proposal.

There may be more misunderstanding and confusion about the conceptual framework than any other part of the research process (or research proposal). Confusion exists on what the conceptual framework is, what its purposes are, and how it is approached. Many economists fail to understand it clearly, so there is little wonder that students of economics are confused. Students typically resist undertaking the writing of a formal conceptual framework unless given no other alternative. Developing the conceptual framework is an integral part of understanding and analyzing the research problem rather than some vague, abstract addendum to the research. The conceptual framework should be understood as an integral component of framing the research problem. The successful development of a conceptual framework helps one understand the problem. In many cases, a clear statement of the problem does not occur until the conceptual framework is developed. It is complementary to the problem description, statement of objectives, and literature review, instead of the fourth step in a series of steps. We treat

127

it as a distinct piece of the proposal in order to provide organization to the proposal, but the reasoning presented in it is intertwined with the previous sections of the proposal.

The conceptual framework is often neglected or hastily formulated, at least in a relative sense, in the research process. One reason for this neglect may be because it requires a somewhat complex set of abilities and characteristics—abstract reasoning, recognition and synthesis of central points, knowledge of existing economic theory, motivation to understand the details of a problem and the surrounding issues. Conceptualizing is also hard work and can be frustrating. Another factor that contributes to the neglect, especially in subject-matter and problem-solving research, is that we may want to get on with the "real research" instead of "theorizing." This attitude is particularly prominent in students, who do not realize that the conceptualizing is an integral part of the research and that failure to conceptualize while *planning* the research may act to slow the research, if not cause fatal errors.

An example may help to illustrate both the importance of developing the conceptual framework and the hazards of not doing it rigorously. Following the U.S. government's breakup of American Telephone and Telegraph (AT&T) in the early 1980s, an editorial appeared in *Newsweek* magazine. The editorial noted that the dissolution of the AT&T monopoly had led to lower long-distance rates but higher local telephone service rates, contrary to the expectation that lower local rates would result from the breakup of the monopoly (Rosenberg, 1983). The writer was correct in his analysis of long-distance rates—the increased competition from other long-distance companies resulted in lower rates. However, his conceptual analysis of the local service rates was erroneous. He failed to recognize that the breakup of the one national company for local service into seven regional companies did not dissolve the monopoly; it replaced the one national monopoly with seven regional monopolies. He apparently assumed, without basis, that size and monopoly power are synonymous. The primary point is that one cannot be assured of a full understanding of the problem without a rigorous conceptual framework.

Previous writings on the conceptual framework tend to be piecemeal and sparse in comparison with writings on the other components of the research proposal. In this chapter, unpublished course notes of Willard Williams (1984) and discussions, debates, and arguments with him on the matter as a student and a colleague are drawn upon heavily. The author's understanding and perspective of the conceptual framework, its purposes, and its value have been influenced more by his thinking on the topic than by any other single individual.

The chapter is organized as follows. The first section defines the conceptual framework. The second section addresses the use of economic theory in the conceptual framework. The third section addresses hypotheses and hypothesis testing, and the fourth offers suggestions on formulating and writing the conceptual framework.

ROLE OF THE CONCEPTUAL FRAMEWORK

Webster's New Collegiate Dictionary (1977) defines *concept* as "something conceived in the mind" and *conceptual* as "of, relating to, or consisting of concepts."
Lacey (1976), in his *Dictionary of Philosophy*, explains that to have a concept of
anything is to be able to think or reason about it; for example, to have a concept of
dog is to be able to think about dogs. Concepts are generalizations.

For use within the confines of economics and economic research, we can
define *concept* as *a logical, mental construction of one or more relationships.*
Several attributes of this definition are important. It is purely mental, is logical,
and can be described; it has been reasoned through sufficiently and presented
with clarity. As such, a concept is inherently abstract (takes some things as given
or assumed). In economics, concepts typically focus on relationships; they may
be basic relationships among variables or more complex systems of relationships.
Also, the relationships are causal in nature. That is, the relationships intend to
explain how or why some things result in (cause) other things.

> In economic research the conceptual framework is a conceptual analysis
> through the problem to all hypotheses relevant to the problem. (Williams, 1984)

Note that the conceptualization is directed toward the problem, not the objectives
or the methods. It is purely conceptual, that is, without regard for empirical evidence or data. Its primary function is to lead to and justify meaningful hypotheses
that are, in turn, subject to testing (verification or rejection).

Most of us conceptualize as a part of our everyday living, although we may
not be aware of it, and therefore the process is probably not analytically sophisticated. The following illustrates the point:

> Who has not entered his automobile only to find that it will not start? With
> presentation of the problem, the conceptualizing process begins. The driver
> begins to reason concerning causes and possible solutions. Since the starter
> works and the engine turns over, the problem is neither the battery nor the
> starter. He might continue reasoning in this manner and, in the process, nar-
> row both the possible causes and alternative solutions. (Williams, 1984)

Grasping the purpose of the conceptual framework is fundamental to using it
effectively. The conceptual framework is not a test of the researcher's knowledge
of existing economic theory. In disciplinary research, it could entail the development or new theory, however. It is not the mathematical derivation of a research
method that is planned to be used in a study unless the intent of the research is to
develop a new method. It is not the explanation of the logic of an econometric
model. It is not the presentation of the standard economic theory that may be
applicable to a given study. Yet the conceptual framework may be, and often is,
related to these other activities. The relatedness to methods and objectives is,

however, through the side effects of the conceptual framework, not as its main purpose or its focus of attention. Diversion of attention to the side effects rather than its function can lead one to overlook or misunderstand important dimensions of the research problem.

The conceptual framework may be viewed as an analysis of the research problem(s) using theory. The theory will probably include economic theory if the research problem is defined by an economist, but it may not necessarily exclude other theory. Theoretical considerations from other disciplines may be appropriate to the conceptual economic analysis. It is most often a process of identifying the appropriate economic and other theory or concepts that are germane to the analysis of the problem, then applying them in a conceptual analysis of that specific problem. Examine the conceptual frameworks in the proposals in appendices A and C. Both address the specific problem identified in the research proposal. That is, the central focus of the conceptual analysis is the main problem issue identified in the specific problem statement. Yet in the case of the proposal in appendix A there is a direct lead-in to the methods and procedures.

If the theory is not already developed, the task may involve formulating theory or refining or modifying existing theory. Theoretical formulations or refinements may be more likely to be required when the research is more oriented to disciplinary matters, as opposed to a heavier emphasis on problem-solving or subject-matter research. This generalization is, however, quite fallible. It is not uncommon for applied research to identify the need for theoretical refinements and result in refinements being developed within the conceptual framework. When that occurs, the research takes on more prominent disciplinary elements.

Although problem-oriented, the conceptual framework often results in other benefits (the side effects mentioned above). One is that it may provide a theoretical link between the objectives and the methods and procedures. The conceptual framework is related to the objectives because the problem leads directly to the objectives. The conceptual analysis thus may help identify relationships, or types of relationships, that are needed to achieve the objectives. It is common for the conceptual framework to point to relevant variables within relationships as well. Williams (1984) characterizes the conceptual framework as an organized "think piece" that, in its analysis of the problem, may include the logic of:

A. Sources of the problem. This may address conditions, circumstances, policies, practices, etc., that cause(d) the problem.
B. Alternative solutions to the problem.
C. Identification of variables relevant to the analysis of the problem.
D. Conceptualized relationships in a system to analyze the problem.
E. Hypotheses to be tested about results of analysis on the problem.

SOURCE MATERIAL FOR THE CONCEPTUAL FRAMEWORK: THEORY

How does one get started on the conceptual framework? Look to the relevant economic and other theories. Which theory is relevant? The relevant theory provides insight into or understanding of the logic of one or more of the five elements just identified, while focusing on the specific researchable problem of the study. This is why a grasp of the theories within a discipline is essential to the conduct of subject-matter and problem-solving research as well as disciplinary research. Placing the problem into a researchable context necessitates its conceptualization which, in turn, requires one to be capable of using the discipline's theoretical tools as a means of thinking about the problem.

Recognizing the relevant theory is only the initial step. The researcher is then faced with *applying* that theory to the specific problem. Ladd (1991) describes this as finding the linkages between the "ideal types" in theory and the "real types" that exist in actuality. If the theory relevant to the problem is well developed in its general form, it will invariably require some adaptation to "fit" the individual research problem. When we conceptualize, we select general ideal types from the body of theory and create our own ideal types within the specific research activity. But when conceptualizing involves only the adaptation of existing theory to a research problem, it is still demanding and requires original thinking. Each research problem is unique, and there are no easy guidelines for conceptualizing.

Consider that the conceptual framework in appendix A adapts standard Marshallian welfare concepts of producer and consumer surplus and international trade concepts to the specific problem of understanding the social costs and benefits resulting from a defined set of domestic commodity and trade policies. In appendix C, the conceptual framework uses elements of merger and investment theory in its conceptual analysis of the expected impacts of mergers in pharmaceuticals on the various stakeholder groups. The conceptual framework in appendix B examines the linkages, flows, and types of impacts caused by the incidence of HIV/AIDS in Russia's macroeconomy. Each of these appendix examples employs a different set of theories as a point of departure, yet each adapts well-understood and documented concepts to its research problem, giving clarity to the understanding of the problem.

For another example, if one is concerned with the subject-matter research problem of water use efficiency of the Nile River in Egypt, it is probable that the general theory of efficiency in the use of limited but replenishable natural resources is relevant to the conceptual analysis of the problem. The theory would be adapted to a particular natural resource (water, and specifically the water of the Nile) in a particular place (Egypt or its subdivisions) in its various uses (agriculture, domestic, industrial). That is, we are not interested in restating the general concepts of resource use efficiency; we can read about those principles elsewhere.

We are interested in understanding the dimensions of the problem of allocation and use efficiency in Egypt. Application and adaptation of the general theory to that situation can enhance our grasp of the problem. The conceptual analysis of the researchable problem may give insight into how to achieve the objectives of the study as well.

Following the same example, it is also likely that Egyptian policies will need to be considered in conceptualizing the nature of the problem. That is, one may need to superimpose water policy conditions or constraints on the theoretical construction of water use efficiency to see how one would expect those conditions to affect efficiency. Egypt's policy of free water for agricultural use affects water use efficiency and may need to be examined or explained conceptually. This also would be part of the conceptual framework.

Sometimes adaptation of existing theory is not sufficient for the analysis of the research problem. In these cases one is left with the choice of either refining the theory or presenting an incomplete conceptual framework. The latter option is obviously less than ideal from a disciplinary perspective. Without advocating this option, it is realistic to recognize that identifying the gap in our conceptual understanding is better than ignoring the problem altogether; that is, the conceptual framework that does not develop the theory makes some contribution by identifying the need for more theoretical development. This alternative, while not the most desirable, may be defensible in some situations—in, say, master's theses— but is less likely to be acceptable in Ph.D. dissertations.

When theoretical refinements are introduced, they are primarily as marginal improvements in the existing theory or adaptation of concepts from other subject-matter areas to a new one. "New theory" rarely, if ever, springs forth in full bloom from the sudden insight of an individual. It is too intricate and logically demanding to occur that way. Instead, it evolves through a process of individuals making sequential, marginal contributions to the broader body of knowledge. Even the "breakthroughs" in economic theory—for example, Alfred Marshall's *Principles of Economics* and John Maynard Keynes's *General Theory*—drew heavily on prior theoretical development.

HYPOTHESES AND HYPOTHESIS TESTING

Recall that a primary function of the conceptual framework is to lead to hypotheses that are relevant to the research project. Hypotheses are the results of the reasoning process embodied in the conceptual framework. They need to be formulated in such a way that they are nontrivial, testable, and capable of being refuted (Ghebremedhin and Tweeten, 1994, p. 47). Recall also that we described a *hypothesis* in chapter 3 as *a tentative assertion that is subject to testing*. As a tentative assertion, it can take the form of a simple proposition of an expected outcome or an assertion of a relationship, or relationships, between or among forces, variables, or events. For example, a simple proposition might be that one production system in a manufacturing operation, based on a particular technology, is more profitable than another production system based on another technology (hypotheses consisting of simple propositions are not limited to problem-solving research projects). An example of a hypothesis of a relationship in, say, a demand study on wine might be that per capita wine consumption in the United States is affected by price of wine, prices of a set of other beverages, per capita income, consumers' religious affiliations, and ethnic background. (See chapter 3 to review the discussion of hypotheses.)

Hypotheses are not always stated explicitly, especially in economic research studies. This may be best understood by considering them in the context of both quantitative (statistical) hypotheses and qualitative or conceptual hypotheses (maintained, diagnostic, and remedial hypotheses) introduced in chapter 3 (Tweeten, 1983). It is not uncommon for even quantitative hypotheses to be left implicit or not formally stated. Consider that the hypotheses in the proposal in appendix A are stated implicitly—as expected outcomes of the policies examined. There is no explicit hypothesis in the proposal in appendix C. In the example of the hypothesis about wine consumption above, researchers often forego the formal statement of the hypothesis regarding the expected relationships. We also often choose not to state the formal hypotheses about expected signs or magnitudes of the parameters in the relationship (e.g., to hypothesize an inverse relationship between wine price and wine consumption, ceteris paribus, or a direct relationship between prices for other beverages and wine consumption, etc.).

There is justification for not stating some hypotheses explicitly, particularly in certain types of research reporting (e.g., disciplinary journal articles) when the hypotheses are already implied. Many economists follow this practice with journal writing and even with research proposals for some audiences. However, we should be cautious with the practice because it carries the risk of compromising the rigor in our analytical thinking. For students developing research proposals for their advisors and advisory committees, a safe guide is that all hypotheses be made explicit; it lends rigor to the process.

Quantitative hypotheses are subject to empirical testing, the most common form being a statistical test. As such, we can identify four general characteristics of hypotheses that facilitate empirical testing (Williams, 1984):

A. *They must be stated in specific terms.* Vague or general statements do not lend themselves to empirical testing.

B. *They require that appropriate data be available or can be generated.* Empirical testing requires quantitative data.

C. *They must be framed so that analytical techniques are available.*

D. *They must have a conceptual basis.* Without the conceptual reasoning, quantification constitutes only association, not causation.

On the matter of qualitative hypotheses in the conceptual framework, it is common for all three types—maintained, diagnostic, and remedial hypotheses—to be used in the conceptual analysis of the problem. Maintained hypotheses often take the form of assumptions we make regarding the conditions in which the problem exists or the relevant dimensions of the problem. When analyzing issues or problems in the commodity production segment of the agricultural sector, for example, we often assume that the segment is competitive (farmers are price takers in both input and product markets, widespread availability of information, etc.), at least in regard to commodities such as grains, soybeans, and cotton. This assumption constitutes a maintained hypothesis. There are no formal quantitative tests for its acceptability, but the "test" is instead by general agreement or disagreement based on logic. Some who cannot accept your assumptions about the framework of the problem as you present it may not accept your analysis of it, thereby rejecting your proposed research altogether. However, some who cannot accept your assumptions will accept your proposed research if the assumptions are defensible.

A point that is especially relevant regarding maintained hypotheses is that the validity of an assumption, and the defensibility of it, often lies in the framing of the problem itself. Consider a research problem that focuses on understanding the forces determining the general level of cotton prices in the United States. We know that in a strict sense the market is not perfectly competitive. Among other things, cotton is not a homogeneous product (it differs in quality among countries, regions within the United States, and even individual bales produced by the same farmer). Yet for a given level of market aggregation, one may treat the market as competitive (use the perfectly competitive market model) for the purposes at hand. This maintained hypothesis (assumption) may be justifiable because the quality differences among countries and regions are stable, the problem involves only understanding the general level of price, and the general price is understood to be only an indicator of what is really a distribution of prices.[1]

Diagnostic hypotheses—propositions about the causes of a problem—have an obvious place in the conceptual framework. The diagnostic hypotheses embodied in the conceptual frameworks of the proposals in appendices A and C are examples. In some instances the conceptual analysis of the problem may begin with a diagnostic hypothesis of the cause of the problem, then proceed to support that hypothesis with theoretical constructions. In this type of case, the hypothesis is either supported by the logical construction of theoretical reasoning or it is not. In other instances, the conceptual analysis of the perceived problem may lead the researcher to a diagnostic hypothesis about the cause(s) of the problem. The theoretical reasoning is essentially the same in both instances, and the test of logical reasoning is the same. The reader, in fact, may not be able to distinguish the researcher's point of shifting from diagnosis of causation to the diagnosis of potential solutions.

Consider, for instance, a research problem concerned with a country's chronic trade deficit. The researcher might present the diagnostic hypothesis that its causes are in part due to monetary policy, in part to fiscal policy, in part to a policy on foreign investment within the country, and in part to domestic food policies that place ceilings on commodity prices within the country. The conceptual framework would, in its theoretical analysis, show why or how each of these policies contributes to the country's trade deficit. (Note that this analysis also would embody maintained hypotheses.)

Remedial hypotheses may be included in the conceptual framework, but are best viewed as optional. As noted in chapter 3, they are closely related to diagnostic hypotheses. Diagnostic hypotheses address the causes of the problem, while remedial hypotheses address solutions.[2] Maintained and diagnostic hypotheses can be tested only by logic of the reasoning as presented. Remedial hypotheses may be evaluated by logic of the reasoning, but they may be presented in such a way as to be empirically testable. Using the previous trade deficit problem example, a hypothesis of the form that a specified combination of monetary, fiscal, investment, and agricultural price policies would reduce or eliminate the trade deficit constitutes a remedial hypothesis. While this hypothesis is testable by weight of logic, it also may be empirically testable via developing an econometric model that captures the effects of these various forces on the country's international trade, gathering data and estimating model parameters, then simulating selected policy change scenarios with it. This approach still relies on logical constructions (of the model), but expected outcomes are empirical (quantified).

Whether hypotheses are quantitative or qualitative, they are an integral part of conceptualizing and the analytical research process. Goode and Hatt (1952, p. 56) state:

> The formulation of the deduction constitutes a hypothesis; if verified, it becomes a part of a future theoretical construction. It is thus clear that the relation between the hypothesis and theory is very close indeed. While it is true that the two can never be satisfactorily separated, it is useful to think of them as two aspects of the way in which science adds to knowledge. Thus a theory states a logical relationship between facts. From this theory other propositions [hypotheses] can be deduced that should be true if the first relationship holds. A hypothesis looks forward. It is a proposition which can be put to a test to determine its validity.

Goode and Hatt proceed to identify three main difficulties people have in developing hypotheses (p. 57). The first is an absence of a clear theoretical framework, or absence of knowledge of theory. The second is the inability to use the framework logically—to see the hypotheses within it. Both difficulties might arise from an insufficient understanding of the body of theory or inability to integrate it into one's reasoning process (these may amount to the same thing). In any event, both difficulties are encountered in both quantitative and qualitative hypotheses. The third difficulty, failure to understand research techniques that allow them to be tested, applies more specifically to quantitative hypotheses.

Ladd (1991) would add a fourth difficulty: that of relating the theoretical hypotheses (in terms of ideal types) to the real types of the research problem.

OPERATIONAL SUGGESTIONS

As you begin to develop your conceptual framework, keep a clear focus on its primary purpose—to analyze the research problem. Its secondary or subsidiary functions will be facilitated by maintaining your attention on the main point of the exercise. Also, keep in mind that the conceptualizing process involves abstractions and abstract reasoning, but applied to the concrete concerns of the issues surrounding the problem(s) when the research is subject-matter or problem-solving focused. Your conceptual analysis will necessarily involve abstractions (simplifications), but with a definite focus—analyzing the problem. Larrabee (1964, p. 171) discusses a "happy medium" that is relevant:

> The happy medium is a constant, cautious shuttling back and forth between words and things. No one can afford to let either one monopolize his time and attention. The "highbrow" is prone to overdo abstractions and to neglect the facts; while the "lowbrow" is just as likely to mock at the theories which he does not understand. To condemn all abstraction and all ambiguity, just because they are subject to certain vices when used in particular ways, is just as silly as to cover them with a blanket of eulogy for their virtues in other contexts. We must never forget that thinking is an art as well as a science; and that different languages may be needed for different purposes. Our policy in regard to abstraction, then, should not be inflexible, but adapted to our purposes.

As you approach the process of conceptualizing, then organizing, a method that is often helpful is to start with simple conceptual models and build on them to the degree of complexity and sophistication that is necessary for the task. This general approach to conceptualizing can be both productive and efficient, irrespective of the target of the conceptualizations. That is, it is useful for conceptualizing when involved in research, policy analysis, and assisting business people, administrators, the public, and others. One must recognize that it is first necessary to define the problem sufficiently to apply some analysis to it. More specifically, there are some steps that may be useful in applying this general guideline:

A. Most importantly, *gain a grasp of the research literature related to your problem.* How have others conceptualized problems similar to yours? How much, or what parts, of those conceptual analyses are adaptable to your conceptual framework? Be aware, however, that you are not likely to find other conceptual frameworks that fit your problem exactly. Modifications will be needed even when similar conceptualizations are available in the literature. You need not reinvent the wheel, but you may need to impose some design modifications to make it fit your wagon.

B. *Reduce the problem to the simplest, most uncluttered set of conditions possible* as a point of departure. Get the problem confined to its most basic dimensions. This is done because the human brain cannot manipulate large numbers of variables or interacting ideas simultaneously (Williams, 1984). You may use assumptions to ignore aspects of the problem that you believe to have little or no bearing on the forces causing it or that might alter its nature. You may even want initially to assume away parts of the problem so as to make it simpler to gain understanding of its dimensions. In either case, these simplifying assumptions (maintained hypotheses) should undergo periodic reexamination throughout the research process (more on this in item D below).

C. *Identify the applicable economic theory.* If the research problem has an international or interregional trade component, then trade theory probably will be applicable. There also could be, for example, reasons for including certain aspects of production theory, welfare theory, and so on. One also may want to identify and apply alternative theoretical formulations to the problem (Castle, 1989, pp. 5–6). There is a range of alternate theoretical formulations from which to draw and little reason to confine oneself to any particular subset or to a doctrinaire "school of thought."

D. *Start with a "base model" analysis.* Using the carefully constructed simplifying assumptions from step A, proceed to construct a "simple" conceptual economic model of the causes of the problem. This may be, for example, a short-run, comparative statics approach to an analysis. It may employ some graphical analysis in the process, although it could proceed on to some more multidimensional or multivariable mathematical analysis. The primary purpose of this step is to get a solid logical beginning on the conceptual analysis. It is unlikely, however, that the conceptual analysis will be complete with this single step, particularly with more sophisticated economic analyses. Most individual research problems require working through multiple successive iterations of conceptualizing before developing something that does justice to understanding the various dimensions of the problem.

E. *Expand the base model analysis to other relevant dimensions of the problem.* Alter and relax restricting assumptions, then reanalyze the dimensions of the problem under modified assumptions. This could, for example, bring in long-run as well as short-run analyses, introduce dynamic aspects to augment the comparative statics, and generally make the conceptualization more complex. Note, however, that the goal is not complexity; it is to get as close to the relevant dimensions of the problem as conditions will permit. There are circumstances when added complexity serves to confuse the issues rather than clarify them.

F. *Assemble relevant, testable hypotheses from the conceptual analysis.* You will have been employing some maintained and diagnostic hypotheses in the process of conducting the conceptual analysis. If your analysis has captured the important dimensions of the problem, adapted the relevant economic theory to analyzing it, and analyzed it effectively, then those expected outcomes should provide a set of testable hypotheses. These may be stated in more general form than formal statistical hypotheses, and they may need to be arranged in some logical order.

There are abundant examples of conceptual frameworks. Economics journals are replete with examples, although they are not always labeled as such; a conceptual framework may be identified with a heading such as model (not to be confused with an empirical model), conceptual model, conceptual considerations, theoretical considerations, or it may simply be integrated into the problem description. In some instances a journal article consists entirely of a conceptual framework (see Smith [1968] for an example). However, the norm for the conceptual framework in a graduate thesis or dissertation is for it to be presented in a separate, formal chapter or section.

The same guidelines for writing covered in previous chapters apply to writing the conceptual framework. Anticipate the writing of the conceptual framework to be rigorous. Expect to write multiple drafts. Seek critical reviews of drafts from persons with demonstrated ability in conceptualizing. Recognize the conceptual framework as a creative and imaginative endeavor (chapter 2) rather than a technical adaptation of theory. Locate and study well-constructed conceptual frameworks.

SUMMARY

The conceptual framework is an often misunderstood component of a research proposal. Its central purpose is to conduct a conceptual analysis of the research problem. It is fundamental to understanding and explaining the research problem. It is not less important in subject-matter and problem-solving research than in disciplinary research. Confusing this primary role with some of its secondary benefits—a theoretical link between objectives and methods and procedures and a foundation for empirical models, for example—is a common problem in constructing conceptual frameworks.

Existing economic theory is the capital stock from which conceptual frameworks are derived, but generalized theory does not, by itself, comprise a conceptual framework (except in the cases of purely theoretical disciplinary research). The conceptual framework entails adaptation of general theories to the specific problem and analysis of the problem using the theory. The adaptation process may require theoretical refinements or developing new theoretical constructions. Most theoretical advancements occur through the need to analyze or explain a problem.

Hypothesis formulation and testing is intertwined with the process of conceptualizing. We are generally familiar with conceptualization resulting in empirically testable (quantitative) hypotheses. Qualitative hypotheses—maintained, diagnostic, and remedial—are also integral components of the conceptual framework, albeit less commonly recognized explicitly. These matters are discussed, along with conditions for quantitative hypotheses.

Some procedural guidelines for formulating and writing the conceptual framework in a research proposal are provided. These include a thorough grounding in the theoretical and other literature related to the problem, identifying relevant economic theory, initiating conceptual analysis with simple formulations and building needed complexity and sophistication on them, and formulating hypotheses directly from the conceptual results.

SUGGESTED EXERCISES

1. Choose and read an article from a recent economics journal. Does it have a distinctly separate conceptual framework (it need not be labeled "Conceptual Framework")? If it does, describe and evaluate it; does it accomplish its intended purpose? If it is not explicit, are elements of the conceptual framework integrated into other parts of the article? Explain. If there is no conceptual framework within the article, explicit or implicit, outline your suggested conceptual framework for the paper.

2. Select a graduate thesis or dissertation and conduct the same analysis of it as for the journal article in exercise 1.

3. Using the two sources from exercises 1 and 2, identify and briefly explain the hypotheses in each. Are the hypotheses explicit or implicit? Address yourself to both quantitative and qualitative hypotheses. Are the hypotheses tested within the article and/or thesis or dissertation?

NOTES

1. While the assumption of a competitive market structure is used to illustrate a maintained hypothesis, competitive markets as a maintained hypothesis is, in the author's opinion, overused by economists, especially in applied empirical estimation research. It appears to be frequently adopted because it facilitates construction of empirical models, not because it has been conceptually defended and justified. *Any* maintained hypothesis made solely for empirical convenience is likely to result in unreliable research results and misinformation.

2. It may be noted that both remedial and diagnostic hypotheses may also be included in the conclusions of a research project. The conclusions, discussed in chapter 10, may lead one to offer hypotheses regarding the causes or solutions to a problem.

9

Methods and Procedures

Distinctions between methods and methodology were discussed in chapter 2, but some clarification of what is meant by "methods and procedures" may be useful. *Methods* refer to the tools or techniques applied in the research process. *Procedures* are the way we put the tools and techniques together, in specific sequences and combinations, to achieve the objectives of the research study. The concern of methodology is that appropriate methods and procedures are selected, designed, and applied so as to both achieve the research objectives and produce reliable knowledge.

In our treatment of methods and procedures in the research process, we will focus largely on methods and procedures for achieving research objectives that contain empirical content. This is distinct from the deductive and inductive methods of reasoning discussed in chapter 4, yet they are related through the process described as the scientific approach. Recall that statistical methods embody inductive reasoning. Economic models, including empirical models, are based largely on deductive reasoning. The methods and procedures for a research project represent a specific application of the scientific approach.

The treatment of methods and procedures in this chapter is from the perspective of planning the project and the written proposal. The reader should recognize, however, that much of what is presented about methods and procedures also applies to the final reporting of the methods and procedures after the completion of a research project. The methods and procedures in the proposal are before the fact, while they are after the fact in the final report. Note that the proposal contains *planned* methods and procedures; there may be reason to alter the plans as the research proceeds. Further discussion of this point is provided in chapter 10.

Organization of this chapter proceeds from a discussion of the purposes of the methods and procedures to a brief historical perspective of empirical methods. The third section covers economic models and modeling because they are such pervasive aspects of empirical economic research. A classification of empirical techniques is then provided. Data considerations are discussed, then the final section offers some suggestions for developing the methods and procedures part of the proposal.

PURPOSES OF METHODS AND PROCEDURES

The central purpose of the methods and procedures in the research project proposal is to provide the plan, and its description, of how the objectives of the study
will be achieved. It is the *what*, *why*, and *how* of the research project: a step-by-
step specification and description of what will be done, how it will be done, and
why it will be done in the specified manner. How it will be done includes the
order or sequence of steps in the process. It includes specification of the quantitative estimation techniques, analytical methods, data that will be collected, and
how the data will be obtained and processed. Methods and procedures also
include how each of these relate to the objectives of the research, how empirical
estimates will be tested and analyzed, model development and design, justification of model mathematical form, and how results will be interpreted.

Refer to the methods and procedures sections of the proposals in appendices
A through C. The methods and procedures for the proposal in appendix A present
alternative approaches for achieving the empirical objectives of the study, with
the final selection dependent on circumstances (which are discussed in the
"Research Methods and Procedures" section). Reviewers were likely less concerned with which specific approach would prevail than with the investigators
having perspective on how generally to proceed and proposing reasonable criteria
for choosing one of the approaches. In the master's thesis proposal in appendix B,
the methods and procedures are specified only in a general way. That is, the types
of tools and techniques are identified, but little detail on their application to the
objectives is provided. While this level of generalization is at the discretion of the
master's student's committee, absence of specifics on methods and procedures is
more common in industry proposals (as opposed to academic proposals), where
evaluators are more willing to leave the analytical details to the "experts." The
Ph.D. dissertation proposal in appendix C contains the analytical details of how
returns from mergers will be estimated, how the differences will be tested, and
how the data for the estimates will be obtained. In this case, the student's committee likely wanted to examine and evaluate the comprehensiveness of her analytical planning.

From a more general perspective, the methods and procedures delineate the
approach for testing the hypotheses of the study. Recall that hypotheses may be
qualitative as well as quantitative and that hypotheses are not always stated
explicitly, particularly in economic research. The methods and procedures specify
the means through which the various hypotheses are tested (examined).

The appropriateness of the methods and procedures depends on the problem
and the objectives. This is yet another reminder of the importance of the problem
identification and specification in the research process. Methods and procedures
directly address the objectives, but the research objectives are derived from the
research problem. Consequently, the research methods and procedures are driven
by the problem and objectives, not the other way around. The caveat for this

"rule" is the instance when the (disciplinary) research objective is to test a new empirical technique, or possibly a new application of an existing technique, in which case the evaluation of the technique is the objective. Even then, the procedures address how to evaluate the technique or its application.

The most complex or sophisticated methods and procedures are not always the most appropriate. Those characteristics may influence the publishability of a piece of research for an academic journal, but the complexity, in itself, will not make them more effective or appropriate. Complexity may only make the methods and procedures inefficient. If you keep your focus on (1) identifying meaningful researchable problems; (2) specifying appropriate objectives; and (3) developing appropriate methods and procedures to achieve those objectives, your research is more likely to be well received. You may not publish everything you do in the *American Economic Review*, but much useful research does not fall within that publication's defined mission.

A HISTORICAL PERSPECTIVE ON EMPIRICAL METHODS

It is interesting that a nineteenth-century economist, John Stuart Mill (1806–1873), formulated the experimental design that has so profoundly affected the laboratory and field sciences and is viewed even today by many in those disciplines as *the* experimental method (Williams, 1984). Mill, who had a talent for classifying and organizing things, provided several classifications of experimental techniques. Two prominent ones were the Method of Agreement and the Method of Difference; these are portrayed in figure 9.1 (for more complete discussion of these methods, see Goode and Hatt, 1952, pp. 74–81; Larrabee, 1964, pp. 236–250).

Mill's Method of Agreement consisted of two parts, the Positive Canon of Agreement and the Negative Canon of Agreement. In the Positive Canon, when multiple occurrences of a given phenomenon have only one condition in common, that condition is regarded as the cause of the phenomenon. Today we object to concluding causation, but instead interpret the experiment as evidence of a relationship between the condition (C) and the phenomenon (X). The Negative Canon of Agreement reasons causation from the negative (absence) of both the condition and the phenomenon; the absence of the condition (C) precludes the existence of the phenomenon (X). As a test of a relationship, the Negative Canon of Agreement is even weaker than the Positive Canon, however.

Mill's Method of Difference is a combination of the Positive and Negative Canons of Agreement. The experiment without the condition (C) (i.e., the control case) results in the absence of the phenomenon (X) while the presence of the condition (i.e., the experimental case) results in the presence of the phenomenon. We reject Mill's conclusion of causation, but his Method of Difference provides stronger evidence of a relationship between C and X than either of the individual approaches in the Method of Agreement.

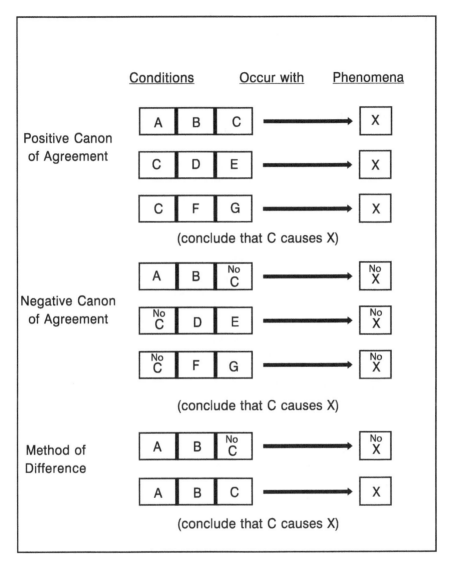

Figure 9.1. Representation of Mill's Methods of Agreement and Difference.

A brief aside on this matter of causation vs. relationship: Experimental methods, whether those as classified by Mill or the statistical and econometric techniques that are used, do not have the capability to establish causation. Our techniques, be they observational or statistical, can establish only *associations*. They help us determine, often probabilistically, whether (and sometimes to what degree) things are related to one another. Evidence of causation is derived by first developing hypotheses of direction of causation from conceptual reasoning (theory), then examining for evidence of the expected relationship. If the empirical evidence exists to support the relationship, then it supports the hypothesis of causation. The causative implications come from the conceptual reasoning (theory) rather than the empirical evidence.

Research in all sciences is concerned with isolating and quantifying effects of individual conditions (variables). In economics, we try to understand and/or quantify the effects of the different variables on a particular economic phenomenon independent of the effects of the other relevant variables. It is important to understand the effect of the price of a given commodity on its consumption independent of (and in conjunction with) the effects of the prices of related commodities, the effects of consumers' incomes, and so forth. Understanding of these relationships is important because we must understand the role of each to understand their collective influence. Thus, the problem of identifying and controlling for other variables that may affect the phenomenon being studied, and identifying interactive influences of those variables on the phenomenon, has demanded prominent attention in all the social and physical sciences.

Control of related variables in the laboratory and field sciences has been attempted largely through variations of Mill's Method of Difference. In economics, control of variables is more difficult because of the complexity of systems and phenomena studied and the infeasibility of controlled experiments in most cases. These factors have made necessary the adoption and adaptation of statistical means of controlling for the effects of variables in economics. Economists were quick to adopt the methods of statistics in the early part of the twentieth century. As the multivariate statistical techniques, particularly regression analysis, became feasible as empirical techniques, a means for statistical "control" of other forces was provided.[1] The field of econometrics grew from this emphasis, which has been fostered not only by the continued developments in mathematical statistics (statistical theory) but also by the technology of calculators and computers. Advancements in computing technology alone have expanded empirical possibilities greatly; we routinely solve algorithms today that were impossible even ten years ago.

MODELS IN ECONOMIC RESEARCH

Economic models are abstractions from reality, developed in whole or part from theory, often expressed in mathematical format, and their purposes are to provide (1) explanations and predictions; (2) discovery; and (3) description and illustration (Edwards, 1978; Ghebremedhin and Tweeten, 1994; Simkin, 1993, p. 67). They can be used with or without recourse to data (i.e., they may be empirical or solely theoretical). When models are constructed with the intent of estimating structure or parameters, the model constitutes a form of hypothesis.

The purpose of a model is to explain how a relationship or system works—to identify the factors or forces that are driving a phenomenon and explain with as much specificity as possible how those forces act and interact to cause the phenomenon. The adaptation of theory to a particular set of phenomena, and for a given purpose, forms a model. If the model explains how a set of phenomena works, it can be used to predict direction of change and identify how policy instruments may be used to affect that change (Edwards, 1978). Although the model structure for the study proposed in appendix A is not specified, when the study was conducted the empirical model drew heavily on the theoretical model of the conceptual framework, yet deviated in several important respects. Some of the deviations were based on data limitations and others on forces believed to have an impact on the policy structure that were not dealt with explicitly within the conceptual framework.

A model may be simple or complex. A simple model might contain a single relationship—for example, an estimated relationship between the price of apartment rents and the apartment vacancy rate. However, a single relationship does not, in itself, constitute a simple model; some single equation models may represent very complex phenomena. Complex models often, however, contain multiple (often many) interrelated relationships or equations. Economic journals abound with examples of complex models.

When we merge data and theory in a model with a particular intent, we build an *empirical model*; this is the focus of econometricians and other model builders. Empirical models may be classified as econometric, optimization, or simulation. *Econometric models* are stochastic (they estimate relationships and parameters with some probability of error) and positive (they estimate relationships and parameters from data as they are produced by the actual working of the phenomena being studied—i.e., they use data produced by the phenomena studied). *Optimization models* (e.g., linear programming) are normative (they derive solutions according to a specified set of objectives—i.e., the specified or "desired" objective function). Optimization models may be nonstochastic or stochastic. When they are stochastic, the probability distributions are derived outside the model and imposed on the optimizing objective function.

Simulation models are mathematical constructs that are positive in their intent (to simulate "reality") and nonstochastic except in cases when simulations are conducted repeatedly with the driving conditions for successive simulations drawn from probability distributions of those conditions. Simulation models, while positive in focus, are different from econometric models because they are inherently artificial constructions as opposed to derivations from observed phenomena. Optimization and simulation models typically involve no statistical hypothesis testing but are often the most effective in cases in which there is no way to observe the phenomena being studied directly. It also should be noted that the types of models are not mutually exclusive. Econometric models, for example, sometimes provide estimates of relationships that, in turn, are used for mathematical simulation.

We economists tend to be fascinated with and devoted to our models, occasionally with a certain amount of blind allegiance that can be detrimental.[2] Models and modeling may embody the best that we, as a field of study, have to offer; and we may even be the best at the modeling activity among all fields of study. It represents a means for us to capture the inner workings of very complex phenomena and understand how and why things happen. It is a means for us to be scientific in the sense of testing theory and a means to provide (conditional) predictions about economic phenomena. But we also should recognize the limitations of models and bear in mind that the models are the means rather than the ends. Everything that is important in economics cannot be modeled; some may argue that the most important things cannot be modeled. Perhaps our most fundamental mistake is to fall prey to the temptation to pretend that something does not exist if it cannot be quantified. Merely capturing thinking within models does not, in itself, legitimize the thinking. As Breimyer (1991, p. 252) states:

> Employment of mathematical methods does not substitute for rigorous theoretical formulations. Rather, it itself requires more precise understanding of the constructs of which the discipline is constituted.

A different but related point is addressed by McCloskey (1990a, pp. 1126–1127). Economists often mistake statistical significance for scientific significance. Regression analysis, while useful, is merely a detail of method.

TYPES OF EMPIRICAL METHODS

Economists use an array of empirical methods or techniques in conducting research and invest considerable attention and energy into choosing, testing, and evaluating these techniques for various applications and into developing or adapting new techniques for applications. The classification of empirical methods presented below is meant as a summary of the general types of techniques, not an exhaustive listing. The classification, while influenced by Williams (1984) and Johnson (1988), reflects the author's own thinking of how the different methods may be grouped. There are warranted differences of opinion on where some specific tools should be classified in relation to others. Nevertheless, the grouping of empirical methods is into two main categories, statistical and econometric tools and operations research tools, plus a method that does not fit well in either category (the descriptive method). Some would argue that the descriptive method is not really an analytical empirical method because it is subjective (interpretive); it is included in the discussion here because it usually entails the application/use of data, albeit in an interpretative way.

The Descriptive Method

This is also called the *historical method*. This amounts to systematic interpretation of historical facts. Examples in economics are numerous. Whenever we present a series of income data, price data, or other data over a period of time and claim it as evidence of a pattern, or suggest evidence of causation based on the pattern, we have engaged the descriptive method. It is because of this common application of the historical method that it is presented as an empirical method. Examining a pattern of, say, exports from a country over a period of fifty years and noting that selected deviations from the pattern are associated with identifiable trade or domestic policy actions also illustrates the descriptive method. This method should not be confused with time-series statistical analyses, however, although there are obvious similarities in purpose. Time-series analyses have mathematical structure and statistical tests that are absent from the historical method.

Economists may use this approach most often in problem definition rather than in the analytical procedures. However, more sophisticated versions of the descriptive method may be implemented in the form of time-series methods, listed below.

Statistical and Econometric Tools

Statistical or econometric tools are all stochastic in nature and are positivistic in their approach. They may be classified as follows.

A. *Simple statistical estimations and determinations of fit and distribution.* These are the estimation procedures and tests that are covered in the early part of our beginning statistics courses (e.g., central tendency, dispersion, distributions, t-test, F-test).

B. *Single equation multivariate statistical analyses.* These include regression and correlation, probit and logit analysis, for example, all in a single equation model framework. Single equation models are included in this grouping, typically estimated with Ordinary Least Squares (OLS) procedures. We also might include in this category related multiple single equation models, estimated with, say, Seemingly Unrelated Regression (SUR).

C. *Structural econometric models.* Systems of simultaneous equations, or simultaneous equations models, are included in this category. They include models of the economy, sector economic and commodity models, and similar structures, and are commonly used in both forecasting and policy analysis. They may be estimated with OLS statistical procedures, but they often use more complex procedures such as Two-Stage Least Squares or Three-Stage Least Squares. Technology in computers and computer software have made estimation/quantification of structural models much easier. This ease of estimation may make the conceptualization of structures all the more important.

D. *Time-series methods/models.* This class of statistical procedures or approaches concentrates on the behavior of (economic) variables or systems of variables through time. Empirical procedures range from simple trend analysis to vector autoregression (VAR). Many economists have held the position that time-series analysis reveals nothing of economic structure, being entirely "data driven," and is of limited use in economics. Others defend these methods as revealing of structure—as, for example, the concept of rational expectations—and as being valuable as a positivistic procedure (see Bessler, 1988, for a synopsis of this view).

Operations Research Tools

This class of procedures includes a wide array of approaches, including static and dynamic, deterministic and probabilistic, simple and complex. The single distinguishing characteristic is that they are all nonstochastic in nature, although some may embody probabilistic characteristics or distributions within them. Some disagree with this characterization because there is a group of programming and simulation procedures that include or embody stochastic information (e.g., stochastic programming and stochastic simulation modeling). These probabilistic elements may be estimated with statistical procedures, but the statistical estimation is external to the operations research procedures, then integrated into the mathematical structures of the operations research approaches. It is on this basis that they are differentiated here, even though the distinction for some purposes is not of great importance. (This seems to be a difficulty encountered with all attempts to group or classify things—the distinctions are blurred at some degree of specificity.) The classification presented places these procedures into two groups: optimization techniques and simulation techniques. For a detailed summary and discussion of the more complex of these procedures, particularly the "stochastic" approaches, see Segarra (1991).

A. *Optimization procedures.* These procedures are normative, not in the sense of advocating what should be, but because they all operate to maximize or minimize some objective function specified in mathematical terms. They are therefore prescriptive in their basic implications.

1. *Linear programming.* This procedure, developed during World War II for the purpose of helping to develop minimum-distance shipping routes for war materials, is so named because it relies on linear input/output relationships. Its integration into economic applications came soon after the war through management applications and is widely used today in research and management applications. It operates by minimizing (usually costs of an operation, firm, or industry segment) or maximizing (usually the net revenues of the same entities) an objective

function subject to a set of resource constraints. Costs and returns are most commonly estimated in budgets, discussed below. Many cost-minimizing transportation and trade models use linear programming.

2. *Nonlinear optimization procedures.* This group includes an array of techniques such as nonlinear programming, risk programming, dynamic programming, stochastic programming, and integer programming in their various specific forms. These procedures are all offshoots of linear programming and all use some form of objective function that is nonlinear, or complicated by other considerations, while the constraint matrix is linear. Use of these procedures has been made possible or fostered by computer technology and software developments.

B. *Simulation techniques.* The tools in this group are positivistic in nature—they attempt to estimate or project events on the basis of "known" behavior rather than solve for a specified objective—but they are nonstochastic. They range from simple procedures to complex mathematical structures. They are represented in three groups.

1. *Budgeting.* This most basic of empirical techniques is the estimation of costs and/or benefits associated with an activity, enterprise, or process. It is an accounting procedure of quantifying the costs and benefits. Budgeting may be comprehensive (account for all costs/benefits) or partial (account for only a portion that is relevant for a specific purpose), and it may represent an actual situation or a hypothetical one. A budget may represent an activity within a firm or represent an entire firm; it may represent a region or country, an economic sector or industry (or some portion), or a commodity, for example. In general, they are simple tools for organizing information, known or estimated, from which costs and returns are estimated. They are not often used by themselves except when they are intended to provide direct estimates of costs and returns to industry or policy makers.

2. *Mathematical simulation.* This consists of a group of techniques that simulate outcomes from nonprobabilistic mathematical structural relationships whose parameters are obtained from information outside the structure. The simplest of these are characterized as economic-engineering models. This procedure essentially links budgets of various parts of an operation together through a set of technical (accounting, engineering, and/or economic) relationships to calculate an outcome. Another technique is the Input/Output Model, made famous by Wassily Leontief and used to quantify impacts among various sectors of national or regional economies. The most complex types of mathematical simulation techniques may be those that Johnson (1988) calls systems analysis. These are systems of equations, often dynamic in structure, that are solved (usually with the aid of a computer) to simulate an outcome.

3. *Probabilistic simulation.* These techniques use information on probabilities as the basis of their simulations. Examples of these procedures include Markov chain models, which use past information on probabilities of changes among defined "states of nature" (stationary or nonstationary transition probabilities) to project changes into the future. Bayesian procedures for maximizing or minimizing expected outcomes and other applications of decision theory and game theory are also examples of probabilistic simulation. Control theory, which uses information on distributions of causal variables (usually a set of them) to solve for decision guidelines to achieve specified objectives, may be the most sophisticated of these techniques.

This classification of methods or techniques is not exhaustive and is not intended to be descriptive. Description and study of empirical methods is the subject of other books and subject matter. But a perspective of the types of techniques is relevant to the study of research methodology. In that context, this section is concluded with a reminder that techniques are not necessarily used in isolation from one another. For example, an econometric model may become a simulation model by using the same structure, with estimated parameters, as a mathematical simulation model; the exogenous variables may be projected or assumed, then the outcomes of the endogenous variables simulated. The procedures of the research project are concerned with how the various techniques, methods, and data are used together to achieve research results.

DATA CONSIDERATIONS

In conducting empirical research in economics, knowledge of data is essential. It is a "practical" aspect of research in general and of economic modeling in particular. Some research activities are not possible because the appropriate data are either unavailable or not feasible. Data are necessary to test concepts or theories empirically, to estimate parameters within relationships, and to establish the applicability of relationships and systems of relationships. To generate effective empirical research, we must know how and where to (1) locate data that already exist; (2) generate data that do not already exist; and (3) determine their reliability and applicability to the research problem at hand.

Secondary Data

Secondary data are available from many different sources, and access to a research library and knowledge of how to use it are essential for data gathering. National governments and international organizations publish data collected through painstaking and expensive activities. In the United States, for example, the Department of Commerce publishes periodic census series, including the

Census of Population, Census of Manufacturers, Census of Agriculture, and numerous others. Other types of statistical series are published by many government departments and agencies—Departments of Labor, Agriculture, Education, Commerce, and others. Many of these data series, while relatively easy to access, are not well known by students until they make an effort to familiarize themselves with the government documents sections of their research libraries. Availability of some data series on agency Web sites is increasing, but the sites are not yet reliable for a complete search for secondary empirical data. However, when the data are available electronically, they usually can be downloaded in easily processable form (compatible with common computer software packages).

National and international organizations, trade associations, and such groups are also productive sources of secondary data. These include groups such as the International Monetary Fund, the United Nations, various commodity and trade organizations, and the United States Chambers of Commerce, to name only a few. One needs to use the reference section of the library to locate these sources, including directories of industry associations and the like. There are also directories of information and data sources such as *Statistics Sources* (Wasserman and Paskar, 1974) and *Business Information Sources* (Daniels, 1985).

Economists, especially those in the industrialized countries where published data are relatively more abundant, may have a tendency to place more faith in the accuracy of data than is warranted. Our empirical research results can be no more reliable than the data with which we work. We should be diligent to remind ourselves that the mere existence of data in numerical form does not, in itself, make it accurate or error-free. We effectively assume that secondary (e.g., published) data are error-free because we have no defense for using it if we do not make that assumption. (The author does not throw stones on this point, being as guilty as any of doing this.) Yet most experienced researchers are well aware of problems with their data in some instances. Some international trade data, for instance, are notoriously poor because of the "political constraints" under which they are collected. It is not unheard of for governments, industries, or firms to manipulate reported data to foster another objective. However, it is unusual to read the findings or conclusions of a research study that, in attempting to explain the existence of some perplexing results, even questioned the reliability of the data. We economists, of all disciplinarians, should know better!

Primary Data

With secondary data, someone else has performed the difficult task of collecting, and hopefully verifying, the reliability of the data. With many research problems, readily available secondary data may not exist. When this occurs, the options are to forego the research or collect the (primary) data needed. Economists avoid primary data collection because it is expensive and time-consuming—we prefer to let other people take care of those "mundane" details. We occasionally become involved with our colleagues in other disciplines in designing experiments, but perhaps not often enough.

Economists' most common means of obtaining primary data is probably through survey questionnaires (personal interviews, mail surveys, telephone surveys). Among the advantages of surveys is that the researcher can develop them to collect data that are specific to the research at hand rather than "making do" with secondary data that happen to be already compiled. That is, one can collect the data to augment the economic model instead of altering the model to deal with secondary data constraints. However, collection of research data using surveys necessitates careful attention to question design and presentation so as to avoid biasing responses. Survey techniques as a method/procedure lie beyond our scope, but some essential points are:

A.   The point, intent, and potential value of the survey should be made clear to those being surveyed.

B.   Surveys should be constrained to the data that are essential to the research; responders will not usually take time to respond to long, complex surveys. Show respect for their time.

C.   Questions should be asked in value-free language, oriented to obtaining (1) factual information or (2) evaluation, opinion, or assessment without connotation of goodness/badness or right/wrong of any particular response.

D.   Pretest the survey to determine if it is being perceived and received as anticipated.

For more detailed suggestions on survey data collection, readers may consult Lansford Publishing Co. (1977, 1980) and Ghebremedhin and Tweeten (1994, pp. 57–58).

A source of data, sometimes overlooked, that may be viewed as primary is data already collected, but not published, from firms and organizations. While these data are often difficult to access because the information may be sensitive or proprietary, in some cases the organizations or firms involved are willing to provide the data *for research use* if appropriate assurances can be provided. The assurances or safeguards usually relate to confidentiality of information about the business or financial affairs of individual persons or firms. Access to this type of data requires an attitude of trust toward the researcher, and this type of relationship requires time and familiarity to evolve. This kind of data access is generally not available to beginning researchers.

Creative Solutions for Data Problems

Although it does not qualify as primary data (or secondary, for that matter), economists have another potential way to deal with some kinds of economic phenomena that are not measured, or measurable, within a system of scalar quantification.

For example, the presence or absence of a phenomenon, be it a cause or an effect, may be captured in the form of a binary indicator (dummy) variable. Or we may estimate different relationships for different time periods, say before and after a particular event, and compare the relationships. We prefer, generally, to work with data in a numerical scalar form. However, the phenomena we study are not always accommodating to that preference, and the information we have to work with is not always available in the preferred form. Creative use of data can surmount some, but not all, problems. On this point, we can probably learn some useful procedures from our sister disciplines, such as sociology and psychology, where data availability may be even more difficult. For example, they use some scaling techniques for measurement of beliefs and attitudes that could also be productive in some economic research.

PROCEDURAL SUGGESTIONS

In concluding this chapter, some suggestions for approaching the methods and procedures portion of a research project proposal are offered. The suggestions are drawn from the author's research experience and from working with students.

A. *Recognize that they are* planned *methods and procedures*. They may be altered as conditions warrant, and the need to change some approaches as the research progresses does not necessarily mean that the plan was not effective or valuable.

B. *Approach the methods and procedures in both general and specific terms.* Look first at the "big picture" of how you will achieve the general objective of the research project—what tools/techniques will be used and how they will be used together, what empirical results will be generated, what analysis will be conducted on the empirical results. Give the reader or evaluator an overview of how the analysis will be conducted and in what sequence of steps. Second, address each specific objective in terms of what will be done, what techniques and procedures will be used, what data will be needed and how it will be obtained, and so on. It may be useful to proceed methodically through each of the specific objectives in this manner.

C. *Be as specific and detailed as possible in the specifications and explanations, but recognize that all specific details cannot be worked out before the analytical work begins.* For example, you may want to present a planned analytical model in its functional form (which variables are hypothesized to impact other variables), but not in its mathematical or structural form (specifically how they are expected to be related), which may not yet be fully developed.

D. *Recognize the audience for which the methods and procedures are written.* For a graduate committee consisting of only economists, standard disciplinary terms and language need not be defined, and analytical techniques that are widely used and understood should not be explained. On the other hand, new techniques may need to be explained and/or justified. When noneconomists are the audience, they may prefer a more general explanation of both standard and new techniques or approaches. With a mixed audience, a more general overview of methods and procedures is recommended, with few specifics for the noneconomist audience, followed by more detail for the economist audience.

SUMMARY

This chapter addresses the methods and procedures as part of the research design and proposal process. It begins by distinguishing between methods and procedures, then considers their purposes. The methods and procedures section of the research proposal has the function of specifying how the research objectives will be achieved. It addresses what will be done, how it will be done, and why it will be done in the proposed manner. It addresses data concerns, use, and availability. It identifies analytical techniques and how data and various analytical tools will be combined to achieve objectives. It also is concerned with testing hypotheses as a part of the process.

A brief historical perspective is given on empirical methods, which date back to John Stuart Mill's Methods of Difference and Agreement. His framework provided the core on which analytical inferences and much experimental design are based. Empirical tools in economics have progressed well beyond his designs into more sophisticated methods. A classification system for empirical methods is provided, with the major categories being (1) the descriptive method; (2) statistical and econometric tools; and (3) operations research tools. These classes have subcategories as well.

The role and function of models is discussed, with models presented as representing our best empirical efforts on one hand and a source of analytical errors on the other. Models can and do help us understand economic structure and complex interrelationships when used well. They can also lead to erroneous conclusions, and even misinformation, when not used well.

All forms of empirical analysis require data, and economists have long-standing concerns about data availability and accuracy. We use data in different ways, often relying on secondary data and sometimes generating primary data. Sources of secondary and primary data are reviewed and guidelines are provided. Some procedural suggestions for organizing and writing the methods and procedures section of the research proposal are also provided.

RECOMMENDED READING

Bessler. "Quantitative Techniques: A Discussion."
Johnson. "Quantitative Techniques."

SUGGESTED EXERCISES

1.   Locate and review two reports of empirical research studies (journal articles or other types of research outlets). For each study, briefly describe how the various empirical methods identified and classified in this chapter were employed.

2.   Using one of the reports from exercise 1, identify the methods and procedures used in the study. Explain the distinction between the methods and procedures and how they relate to one another.

3.   Obtain a research proposal written by a student and critique the methods and procedures section of the proposal. Be specific in your comments and provide suggestions for improvement of weaknesses in the proposed methods and procedures.

NOTES

1.   It may be noted that many of the early applications of statistical methods took place in the then emerging field of agricultural economics, with its problem-solving interest and tradition of and penchant for empiricism. Agricultural economists also led in the use of computers and techniques such as linear programming after the Second World War. According to McCloskey (1990a, p. 1124), "Agricultural economics invented econometrics."

2.   Those who can appreciate a humorous and critical treatment of modeling in economics may want to read Levins (1992).

III

Closure of the Research Process

10

Reporting the Research

Those among us who are unwilling to expose their ideas to the hazard of refutation do not take part in the scientific game.

Karl Popper

Chapters 5 through 9 focused largely on the research proposal. Relatively little attention has been devoted to carrying through with the analysis—the specifics of the analytical phase of each research project. The reason for this omission is that details of analytical methods are not a part of our objectives in the study of research methodology and you have studied details of analytical methods in other contexts, settings, books, and courses. The last step or phase in the research process, after the plan has been implemented, is reporting the research activity. This matter often receives little formal attention and is the topic of this chapter.

Reporting the research involves more than just providing the results of the process. It also includes describing and explaining how the results were obtained and rationalizing the way they were obtained (by implication, the defense of the process). Recall that the validity of the research results (which are the realization of the research objectives, which are, in turn, directed at the research problem) depends on the methods and procedures, not on the results obtained. The manner in which the findings and conclusions were derived is equally important with the findings and conclusions that were obtained. All aspects of the research process are important in reporting the research.

This chapter is presented as follows. Considerations for choosing or selecting types of research reports (outlets) are discussed, then components of research reports are presented and discussed. Suggestions are given for writing a report's methods/procedures and findings as well as its conclusions. The chapter concludes with some observations and guidelines on two aspects of research reporting that are often troublesome: publishing and authorship. Before embarking on this chapter, however, the reader is advised to reread the sections on "Importance of Writing" and "Writing Guidelines and Tips" in chapter 5.

159

TYPES OF REPORTS

Research is reported in many different forms and formats. It is reported in journal articles (disciplinary, interdisciplinary, multidisciplinary, and/or subject-matter journals), in all sorts of technical research reports (such as various reports published through research organizations), in monographs or books (some of which are essentially extended research reports), or in graduate theses and dissertations. Research is also reported orally at different types of professional meetings, seminars, symposia, and workshops sponsored by disciplinary or professional organizations, government groups, industry groups, and so on. However, since oral reporting of research is practically always based on an already written version of the research, attention in this chapter is devoted primarily to written versions of reporting research.

Journal articles are the most condensed form in which research is reported (note that journal articles are not confined to reporting research; they also may be used to elucidate and discuss current issues and concerns, review historical patterns and trends, etc.).[1] Journal articles are terse in their format of reporting research and typically concerned with space considerations (number of pages). Journal articles are also the most prestigious and "desirable" form of reporting disciplinary research, especially within universities and other types of research institutions. The primary reason for this status for journal articles is, in the author's view, due to the somewhat extreme importance placed on journal publishing by administrations of most universities (where most economics researchers are employed). That importance is based largely on two considerations: (1) journals have a peer-review process for evaluating the quality, relevance, contribution, and/or importance of a written paper; and (2) journals typically are believed to obtain the widest distribution of all types of research reports among groups of disciplinary and subject-matter readers.

It is tempting to elaborate personal views on the problems with journal reviews and publishing and balance between types of published research reports. However, it would not serve the objectives at hand. It will suffice to point out that economics demonstrates, as a fundamental principle, that efficient or optimal positions or solutions rarely, if ever, lie at the extremes; efficiency dictates optimizing at the margin. If that generalization is valid for publishing/reporting of research, placing all the emphasis on journal articles, or on any other type of research reporting outlet, is a suboptimal solution.

When we report research through journal articles, we are providing the results almost exclusively to a group of disciplinary and/or subject-matter specialists. The articles are written compactly enough that few persons outside a trained, specialized group can read the paper with confidence, and it is rare for anyone else to even try. To reach other audiences other types of outlets must usually be used. The limited audience represents a disadvantage for disseminating research information through journal articles. Its offsetting advantage is that it is a relatively effective and efficient means of communicating within a group with specialized, focused research interests.

Another limitation of research dissemination through journal articles is that many research projects and processes are too involved, complex, and intricate to be effectively communicated, even to specialists, within the confines of a 20-page journal article. There are cases in which reporting of research, or doing justice to its reporting, cannot be done within the limitations of journal article format. In these instances, the more appropriate research report outlet may be one that facilitates more detail and more explanation. Examples of these are technical research reports such as those published by the National Bureau of Economic Research, research institutes, and government and international agencies. There are also monographs, technical bulletins, and papers published by, for example, the Federal Reserve Banks.

On the other extreme of length and completeness of the research report is the graduate thesis or dissertation. This constitutes the report of the student's research for his or her graduate committee that is intended, among other purposes, to demonstrate the student's command of the topic, level of research achievement, and final qualification to receive an advanced degree. In this case, the writer must ensure that all things done and found are clear to the committee members—not just the whats, but the hows and whys as well. Graduate student research papers often are longer or more wordy than necessary, but they do not often contain more information than is needed for committee members to grasp the research. However, completeness rather than efficiency is preferred in graduate students' papers if both cannot be obtained. Put another way, in writing graduate research reports, completeness is first priority and efficiency is second (one might argue that these are reversed with journal articles for reporting research). Once completeness is achieved, efficiency in communication can be improved with editorial refinements.

In completeness and clarity of research reporting, outlet forms such as technical reports, bulletins, and monographs lie between the two extremes of journal articles and theses/dissertations. Each report is written with a style and a level of complexity consistent with the interests, backgrounds, and technical capabilities of the targeted audience. For example, research reports written for policy makers usually are better written in nontechnical terms, and usually are less technical the higher the level of the policy makers. On the extreme, they may have to be written in a style similar to that of popular (journalistic) writing, although with a more organized progression of thought. Sometimes different reports from a single research project or activity may be written for different audiences. Many researchers avoid rewriting for different audiences. However, this type of rewriting activity may, in the broader sense, be more productive than plunging ahead to another research project, especially when the research has both disciplinary and subject-matter or problem-solving components. For example, the output of a given research project may be written as a journal paper with emphasis on the conceptual or procedural aspects of the research for a disciplinary or subject-matter audience and as a technical bulletin with emphasis on the findings for a policy or industry (problem-solving) audience.

COMPONENTS OF THE RESEARCH REPORT

A stylized version of the components of a research report are: title, acknowledgments, abstract, table of contents, introduction, literature review, conceptual framework, methods and procedures, findings, summary and conclusions, references, and appendices. You may note the overlap between this list of components and the components of the project proposal from chapter 5. In fact, some sections of the project proposal, if done well initially, may comprise sections of the final research report with only minor rewriting. These similarities and differences will be noted as each component is considered in more detail.

Note also that the listed components relate more closely to the research report as a thesis or dissertation than as a journal paper. Some components may not be a separate section in a journal paper. However, the components (but not necessarily the headings or sections) are standard across types of reports, except for the table of contents, although relative emphasis may vary considerably.

A.   *Title.* The same guidelines apply as for the title in the research proposal; in many cases they will be the same. Often in more formal reports, the title is on a title page that also includes author(s), affiliation(s), key words, and similar information. Thus, the title page may contain information identifying different aspects of the project proposal.

B.   *Acknowledgments.* Recognition of assistance and support by individuals and organizations is a detail that should not be overlooked (more is said about this in the last section of this chapter). One should recognize the assistance and contributions of individuals who have been helpful in various capacities—ideas, data, computer assistance, and writing. One should likewise acknowledge financial and other resource support by organizations. Acknowledgments in research writing are usually short and direct, without elaboration. Their placement and form vary. For example, journal articles typically place them in a footnote. Theses and dissertations conventionally have an acknowledgements page following the title page.

C.   *Abstract.* The abstract is a very compact summary of the research report. Its length typically varies from fifty words to two or three typed double-spaced pages. You may sometimes see it called an "executive summary." Although brief, the abstract is extremely important because it assists readers in determining their interest in various parts of the report. Because it is brief, writing it is usually a demanding task.

D.   *Table of contents.* This is a listing, or outline, of the organization of the report. It shows headings, subheadings, and other divisions and components. The table of contents is often but not always included in formal research reports (theses and dissertations). Journal papers delete them in most cases. An extension of the table of contents in the form of a list of tables and/or a list of figures is sometimes included as well. Lists of tables are presented here as an extension of the table of contents, instead of as separate components, because they perform a similar function—to help the reader follow the organization of the report and to locate sections or segments of it.

E.   *Introduction.* Editorial style for introductions varies. Some reports, usually with longer introductions, have a section with an "Introduction" subheading while others have no subheading, particularly when the introduction is brief. It is common for graduate papers to have a separate introduction section (chapter).

As the introduction to the research report, this section relies on the thoughts, reasoning, and writing embodied in the problem statements and statement of objectives in the research project proposal. For example, the problem statements from the proposal in appendix C were included in the introductory chapter of the resulting dissertation. A typical difference is that the introduction is usually written in more of a narrative format, whereas the problem statements in the proposal may be less explanatory.

The purpose of the introduction is to prepare the reader to digest and understand the research and the research report. Thus, readers must understand the problem and objectives. They often need more than those elements in the introduction, however. An overview of the methods and procedures may be included, for example, as well as other material, such as descriptions of study areas or industries, institutional or policy settings, and similar matters. A compacted version of the literature review (rather than a separate literature review), or a portion of the literature review, may be included in the introduction for some purposes and/or audiences.

F.   *Review of literature.* The literature review serves the same purpose in the research report as in the research proposal. It may be organized and presented in the same way, with the same organization. The literature review in the proposal in appendix C, for example, was essentially the same in the resulting dissertation. A difference is that it often contains more material than does the proposal because it is not uncommon for additional relevant prior research to be found during the course of conducting the study; this new material is summarized and added. Remember that the description of the related literature must be designed for the audience. A graduate committee may want much more detail than an industry group, for example, and a journal will want little more than references to the relevant literature in some cases.

G.   *Conceptual framework.* The conceptual framework likewise serves the same function in the research report as in the proposal. Under ideal circumstances it can simply be inserted into the research report. However, it usually undergoes some refinement in the process. A formal conceptual framework is not a separate part of all research reports, sometimes being integrated into the introduction and/or methods and procedures. Alternatively, portions of literature review may be integrated with the conceptual framework. Much of the conceptual framework may be omitted in most instances when the readers are expected to be noneconomists.

H. *Methods and procedures*. The methods and procedures section from the project proposal provides less assistance in writing the research report than do the prior sections of the project proposal. The methods and procedures in the proposal are written before the fact while they are after the fact in the research report. Williams (1984) calls this the "show and tell" section. This part of the report explains how the analysis portion of the research was conducted—what, why, and how. It includes description of data, data collection and manipulation, data sources, analytical procedures, including models developed and used, empirical procedures and techniques, and analyses conducted on the empirical results. Problems encountered in the process of conducting the research and the manner in which the problems were addressed and/or resolved may be described.

I. *Findings*. This component of the report presents *and explains* the results of the analysis. This is the end product of all the analytical procedures from which the objectives are either achieved or not achieved. Hypotheses, both quantitative and qualitative, have been tested; the results (findings) of that process are reported in the findings section. When the research has empirical context, mere presentation of empirical results is not sufficient. Analysis and interpretation of those results are needed to derive the information of the findings. For example, estimation of a demand relationship and its parameters is merely the first step in the process of deriving findings. Subsequent steps that are often needed are calculation of price, cross-price, and income demand elasticities and examination of what those elasticities mean. Such findings may need to examine income and substitution effects from changes in product prices and industry income effects from changes in consumers' incomes. The methods and procedures provide explanation of how the findings were generated; the findings provide the information on what was found, including an *explanation* of it.

J. *Summary and conclusions*. The summary is most often presented as an overview of the entire study, with emphasis on problem(s), objectives, methods and procedures, and findings. Its purpose is to provide the reader with a general understanding of the research project. Summaries should give the reader an accurate general understanding of the research effort, yet be brief. Even with long research reports such as theses or dissertations, summaries of two to six pages are often sufficient. The conclusions, on the other hand, comprise the researcher's interpretations of the findings. They extend beyond the strict boundaries of what was found through the analytical procedures to the *implications* of those findings. The researcher takes the initiative to provide readers with his or her insight gained from conducting the research. Conclusions will be discussed in more detail in a subsequent section of this chapter.

K. *List of references*. The list of references in the final report serves the same function, and may follow the same style and format, as references in the project proposal. It is a listing of the references used in all parts of the research report.

    L. *Appendices.* Appendices may be very useful for organizing the material in a research report. They are used to present material that might disrupt the flow of thought in the body of the report or material in which only a portion of the audience has an interest. For example, certain theoretical digressions or particularly complex mathematical proofs or derivations may be better placed in an appendix. Sometimes the details of results of statistical estimations or tests may be presented more effectively in an appendix than in the body of the report. Findings in other forms—tables or graphs—or "intermediate" findings (results of analyses that were generated before the results that directly addressed the stated objectives) may be suitable for presentation in an appendix. Which material, if any, to place in an appendix is a matter of judgment. The guideline is placement and organization of material as it appears most efficient to you, then reliance on thorough, critical reviews by colleagues. As always, try to select reviewers with representation among those with demonstrated writing achievements and an understanding of the audience for which the report is being written.

WRITING THE METHODS/PROCEDURES AND FINDINGS

The purpose of this section is to provide more detailed suggestions on the writing of these two sections of the research report; they depart from the content of the research proposal in important and significant ways. Some suggestions and perspective on the methods and procedures are covered first, followed by more discussion of writing the findings, then consideration of the delineation between them.

    Recognize that the methods and procedures are written largely for other economists. This does not say that noneconomists are never interested in the methods and procedures of your research, but those outside the discipline will tend to leave the judgments and assessments of validity of specific methods and procedures to the disciplinarians. However, a well-written, thorough description of methods and procedures is necessary for persons within one's own discipline to evaluate the research. While the purpose is to explain what, why, and how, the effectiveness of the written description depends on two characteristics: organization and thoroughness (note that the need for thoroughness may vary with the type of report, as discussed previously). Reconstruct or describe the procedures in a straightforward, logical sequence. Start the section with an overview description of steps in carrying out the analysis, then proceed to describe each step or component of the analysis.

    As you were implementing the research plan, you should have kept records and notes to help in the reconstruction of the methods and procedures. Do not trust details to memory because there will usually be too many for mental recall without some assistance. A daily log or journal of your activities, decisions, and steps is a well-advised organizational technique; it helps to explain the sequence and rationale of the procedures. Explain the data used and the source(s) of them, and any manipulation of or adjustment to the data. Explain the model(s) used and be explicit about any assumptions made. Explain, rationalize, and/or justify your analytical assumptions. You also may explain problems encountered and the manner in which they were resolved. Likewise, note unsuccessful approaches, techniques, and procedures; it may help others avoid problems or mistakes. You may need to explain empirical techniques, for example, if they include procedures that are not generally understood within the disciplinary, subject-matter or problem-solving group interested in the research. Remember your audience when deciding the extent and type of technical detail to include. Explaining details of techniques that are widely understood, such as assumptions or inner workings of standard statistical procedures, is not recommended for a disciplinary audience. Newer, less commonly used tools and techniques may need some explanation beyond referencing their source, however.

Be sure that readers can understand how calculations were made and estimates were derived. Make sure all variables are defined, including units of measurement. This kind of detail is easily overlooked, but the meaning of a parameter cannot be determined unless the units of the variable are defined. Do not hesitate to use graphs or other aids where useful and appropriate. For example, a general schematic of a multisector model may make the structure clear and provide the reader with the perspective needed to grasp the equations and linkages within the model (see appendix B). Tables may also be useful in the methods and procedures, particularly for data description and/or intermediate results. Note, however, that long, detailed tables of this type may be better placed in an appendix.

The findings revolve around the results of the analytical process that the objectives identified to be determined. The findings section presents and explains the results of the study. They return to the hypotheses of the study and present the tests of their validity; this includes both the quantitative and qualitative hypotheses. Merely presenting empirical estimates, for example, is not sufficient. Those estimates must then be analyzed, interpreted, and possibly tested to make the findings complete. They may need to be evaluated considering theory and/or prior empirical research. Students conducting research that has a substantial empirical content often assume that the empirical estimation portion of the research constitutes the findings. In reality, the empirical results are frequently only the beginning of the meaningful part of the research. Competent technicians can produce the empirical estimates, but economic understanding, expertise, and insight is required to analyze the meanings and implications of the estimates (i.e., complete the analysis).

This distinction between empirical results and the completion into relevant findings is the essence of McCloskey's (1990a) distinction between statistical and scientific significance. The full presentation of findings must proceed into the significance of the analytical process. Findings cannot stop with merely knowing the magnitude and statistical significance of estimated parameters, for example. They must also delve into the ramifications of the parameters—what they tell us about the economic phenomena we are studying.

Tables and figures are often effective tools for presenting findings; they help to organize and synthesize the information in the findings. They also provide focal points around which to organize the necessary explanation, discussion, and evaluation of results. A recommended approach is to construct the tables and figures that form the core of the findings first, then use the narrative (explanation and discussion) to augment the tables and figures—use the tables/figures to support the narrative, and vice versa. (See chapter 5 for style considerations for tables, charts, and graphs.)

Sometimes the separation of methods and procedures from findings is not clear. There is no need to deal with *how* results were obtained in the findings, but in some instances the most effective way to explain how results were generated is to provide the result as a part of the methods and procedures. A useful guideline is to present only the results that directly achieve the objectives in the findings section of the research report. All interim or intermediate results generated in the process of obtaining those results are often better presented in the methods and procedures. This guideline is not infallible, however, and the final decision should always consider the order and logic of presentation that communicates best with the targeted audience. Critical reviews by peers and advisors are valuable in making those decisions.

WRITING THE CONCLUSIONS

The most common mistake with the conclusions section of a research report may be that people confuse findings and conclusions. *A restatement or summary of findings does not constitute conclusions.* One way to distinguish between them is that the findings may address and test hypotheses, including qualitative hypotheses (such as whether a model construction appears to represent the phenomena being modeled), while conclusions are concerned with the implications of tests of those hypotheses. The conclusions address the question of "So what?" regarding the findings. Conclusions extrapolate beyond the findings. Drawing conclusions involves examining and interpreting the implications of the study, sometimes beyond the strict boundaries of the study. Conclusions are understood to allow the *judgment* of the researcher(s) about matters related to the problem addressed by the research. This judgment must be supported by logic (have a logical foundation that is clear), but the evidence is not necessarily irrefutable. Williams (1984, p. 81) states:

> Conclusions are a final *inductive* phase of the research. There is no fixed set
> of rules for inductive reasoning, as for deductive logic. The perimeters of

inductive reasoning are a matter of judgement. In this regard, it generally is
assumed that it is wiser to understate than to overstate.

More simply, Brorsen (1987, p. 319) describes conclusions as what was learned
from a research process.

Conclusions may, for example, offer insight regarding the implications of the
similarities and/or differences between the findings of your study and findings of
related studies. They may identify perceived shortcomings of your study, suggestions for improvement, and/or recommendations for further research. They may
delve into the policy implications of the findings, even though policy analysis was
not necessarily a part of the objectives. They may identify unanswered questions
or unresolved problems. They may examine the extent to which the implications
extend beyond the study area, economic sector, or other boundaries of the specific
study. Conclusions may take the initiative to specify what the study *does not*
imply, as well as what it does imply. That is, the researcher may see a need to
ward off potential improper use of the research results. Conclusions may address
any aspect of the research, such as the problem, objectives, methods and procedures, and/or theory.

Most of us find conclusions difficult to write because they require reflection
and a different sort of creative thinking and writing. You have to shift your mental processes from the more rigorous mind-set of "justify and defend each step of
the procedure" to the broader, more inductive mind-set discussed in the quotation above. We usually require time to reflect on the implications of findings and
mentally adjust before writing conclusions. One approach for achieving this
mental adjustment is to first write the summary, then set the report aside for several days and possibly discuss the research with colleagues before attempting to
write conclusions.

PUBLISHING

The written research report is the most powerful means of communicating knowledge within the research and scientific community. Publication of the products of
the research process is a matter of importance for several reasons. From a broad
scientific or disciplinary perspective, publication represents the primary means of
disseminating the products of research within the scientific research community.
If scientists did not communicate, progress in developing new reliable knowledge
would be very slow, even "grind to a halt" (Ayala et al., 1989, p. 10). Additionally,
the review process associated with publication within the scientific community,
and the subsequent process of the persons within a discipline or field of study
assessing the research and its value, is a part of the basic mechanism or evaluating
the research and its value to science. For an individual within a discipline, publications are the means for disseminating one's research knowledge and findings to

others. They are also a primary means of establishing a research reputation and building a career.

Conventions in publishing are also the primary means for documenting researchers' achievements and ideas. Ayala et al. (1989) point out that recognition by one's peers is an important motivating force in research and the advancement of science, although the motivation for personal credit can in extreme instances be counterproductive. Although once published, research results become public property, scientific ethics dictates that the original idea (and its originator) be recognized. Ayala et al. (p. 9) explain how the "system" evolved:

> The system of associating scientific priority with publication took shape during the seventeenth century in the early years of modern science. Even then, a tension existed between the need of scientists to have access to other findings and a desire to keep work secret so that others would not claim it as their own. Scientists of the time, including Isaac Newton, were loathe to convey news of their discoveries to scientific societies for fear that someone else would claim priority, a fear that was frequently realized. The solution to the problem of making new discoveries public while assuring their authors credit was worked out by Henry Oldenburg, the secretary of the Royal Society of London. He won over scientists by guaranteeing rapid publication in the Philosophical Transactions of the society as well as the official support of the society in case the author's priority was brought into question. Thus, it was originally the need to ensure open communication in science that gave rise to the convention that the first to publish a view or a finding, not the first to discover it, gets credit for the discovery.

While publication in scientific journals is the most prestigious outlet for the products of research, it is by no means the only outlet. Others include research or technical bulletins and reports, proceedings papers, symposia and workshop papers, and working papers. In some cases, these other types of outlets provide an opportunity to offer research results to the appropriate scientific group, obtain comments and suggestions, then refine the research before proceeding to a journal submission. In other cases, these other outlets are the most effective form of the final research report. The Committee on Science, Engineering, and Public Policy (1995, p. 10) states:

> Publication in a peer-reviewed journal remains the standard means of disseminating scientific results, but other methods of communication are subtly altering how scientists divulge and receive information. Posters, abstracts, lectures at professional gatherings, and proceedings volumes are being used more often to present preliminary results before full review.

Whatever the final form of the research report, however, the role of peer and advisor input into the process is essential and its importance cannot be overemphasized. Take appropriate precautions that your published research products are both defensible and clearly presented. Among the common reasons that proposed refereed publications fail to be accepted are the following (Brorsen, 1987, pp. 316–317):

1. Inadequate identification of a research problem.
2. Inappropriate or unclear methods and procedures.
3. Inappropriate material for the proposed publication.
4. Failure to communicate what is important and original.
5. Poor organization.

Also, results and/or analytical approaches that deviate from accepted norms or conventional wisdom will be held to a higher standard (Brorsen, 1987, p. 316).

AUTHORSHIP

Recognition of contributions to research and science—the allocation of credit—is a sensitive matter and deserves thoughtful attention. In Merton's (1988, p. 621) words:

> While the lay reader located outside the domain of science and scholarship may regard the lowly footnote or the remote endnote as a dispensable nuisance, these are in truth central to the incentive system and an underlying sense of distributive justice that do much to energize the advancement of knowledge.

Credit in research papers is recognized in three basic forms: citations or references, authorship, and acknowledgments. Citations and referencing were addressed in chapter 7. The primary concern here is with the decisions about and handling of authorship, with some attention to the alternative of acknowledgments in giving credit for contributions to research. Readers are referred to Ayala et al. (1989) and the Committee on Science, Engineering, and Public Policy (1995) for excellent perspectives on these matters.

A guideline for authorship is that all persons directly involved in the research and making substantive contributions should be included in the authorship. The vagueness of the guideline lies in the subjectivity of what constitutes "direct" involvement and/or "substantive" contributions. Some examples may help. Direct involvement and making substantive contributions means being involved in the planning and conducting of the research in ways that affect its overall approach and outcome. If a person is involved in a research project in a decision role of shaping its perspective and design, that person should be at least considered for inclusion in authorship.[2] If, on the other hand, the individual's contribution consists of providing information and advice on specific, technical points, then that more limited, advisory role may be more properly recognized in an acknowledgment (often included in a footnote or short section somewhere in the research report). In some circumstances in which more than one report of the research is published, relative contribution may be appropriately recognized by inclusion of contributions as coauthors on some reports but not others. Also, authors can be listed in different order on different reports. Normally, assistance in the form of secretarial functions, student assistants' compilations of data, technical help with computers, and similar activities are adequately credited with acknowledgments.

Failure to extend credit appropriately may carry no immediate repercussions, but it is likely to offend someone's sense of equity and cause problems with assistance and cooperation in future research efforts. This problem may exist both within and across disciplines, but its consequences may be more severe in cooperative or collaborative research that crosses disciplines. We will digress briefly on this point.

Problems with extending credit may often be more troublesome with multidisciplinary collaborative research efforts such as between economists and researchers in the physical or biological sciences. This may be attributable to some basic differences in views about what constitutes research and research output. It is not uncommon, for example, for researchers in the laboratory or field sciences to view the generation of *data* as the end product of research, while economists are likely to see those data as input for their research process. Those laboratory or field researchers are apt to believe that they deserve coauthorship when an economist uses their data. The economist, viewing the published data as being in the public domain, may see the proper credit as an acknowledgment and appropriate reference to the published data.

This type of complaint about economists' use of research from other disciplines arises in several contexts and settings. Yet a related problem sometimes occurs with the complaint in the opposite direction, which may also be attributed to differences in perceptions. Laboratory and field researchers are prone to assume that any activity that does not involve the generation of primary data (i.e., logical positivism only) does not qualify as research and may not deserve even an acknowledgment. A real situation illustrates this problem.

An economist served on the committees of two graduate students in another discipline who became interested in comparing the cost of a new procedure for conducting a particular natural resource enhancement practice. Being encouraged by their interest in an economic component of their research, he willingly provided advice and guidance as requested. The substantive assistance he provided was primarily on how to conduct cost calculations and allocations that were technically correct and could be defended; the students were making severe errors in several areas until he redirected their procedures. Their results were eventually published in a disciplinary (noneconomic) journal. In this instance, the economist's contribution to the substance of their analysis exceeded that of the major professor. Yet their major professor was included as a coauthor while the economist's assistance was not acknowledged. The point: Concerns about professional courtesy, equity, recognition, and credit extend in all directions.

Conventions or protocol on order of authorship differ across disciplines and organizations. Some place great emphasis on order of listing of authorship while others place little. Some organizations even dictate that research administrators or supervisors be given coauthorship irrespective of the extent of their involvement in the research. This is poor policy because (1) it may violate researchers' sense of equity and tends to be a disincentive in the research process; and (2) it fails to recognize the connection between responsibility for research quality and authorship. If a researcher's name is on a publication, he or she must be prepared to share responsibility for all the research findings and conclusions. This does not suggest that research supervisors should never share authorship, but rather that authorship in those relationships should be determined by the extent of involvement in the substance of the research. There are limited instances in which, for example, supervisors contribute nothing to the research beyond some editorial suggestions on a draft of the final report (and some cases when they do not even read the manuscript), but are listed as an author. These are the cases that are obviously inappropriate.

To avoid problems with authorship and acknowledgments, and the division between them, the safest, and recommended, approach is to discuss the matter openly among the group conducting the research in question. Division of the responsibility between the planning, conceptual, procedural, empirical, and written contributions to the overall effort and its effectiveness must be considered. For example, when other contributions are approximately equally shared, first authorship usually goes to the person carrying the major responsibility in producing the written report of the research. But all aspects of the research effort should be considered. You will seldom make serious mistakes on these matters if you will discuss authorship openly and frankly. This applies equally to authorship between faculty and students, senior and junior researchers, researchers in different disciplines and in different organizations, and peers.

SUMMARY

This final chapter addresses the final step in the research process—writing the research report. The report may take different forms, including journal articles, theses and dissertations, research bulletins and monographs, with journal articles representing the most compact form and theses and dissertations the most exhaustive. Whatever form the reporting takes, the "standard" list of components includes title, acknowledgments, abstract, contents, introduction, literature review, conceptual framework, methods and procedures, findings, summary and conclusions, references, and appendices. Each component is not necessarily an explicit portion of every research report (the report is tailored to the expected audience), and each component is not always a separate section in the written report.

Suggestions and guidelines on writing the report are provided, with emphasis on the methods and procedures, findings, and summary and conclusions. These are the components of the final report that differ most from the research proposal. Methods and procedures in the report are written in greater detail, and after the research is completed. Findings are unknown until the research analysis is completed. Conclusions represent a final, inductive phase of the analysis, which often requires the researcher to adjust his or her mental focus.

Considerations on publishing, recognition of research contributions, and authorship are discussed. These matters relate to recognition of intellectual and scientific contribution. They are matters of academic and scientific concern because they affect individual motivation and disciplinary progress. The primary vehicle for receiving credit for one's contributions is through authorship, citations, and acknowledgments. Some recommendations for dealing with these matters are offered.

RECOMMENDED READING

Alley. *The Craft of Scientific Writing*, pp. 131–182.
Committee on Science, Engineering, and Public Policy. *On Being a Scientist*, pp. 9–20.
Hamermesh. "The Young Economist's Guide to Professional Etiquette."
Thomson. "The Young Person's Guide to Writing Economic Theory."

SUGGESTED EXERCISES

1.   Select two research reports of different types (e.g., journal article, technical monograph, thesis, etc.). Examine the content and organization of each and evaluate their similarities and differences. Do both contain all of the components of a research report identified in this chapter? Is each component explicit in each report? Do the two reports place different emphasis on the various components?

2.   Using the two reports from exercise 1, identify the target audience, in your opinion, of each report. For each report, identify an alternative audience and describe how you would rewrite or redirect the research report for that audience.

NOTES

1.   Helpful suggestions for preparing papers for economics journals are provided by Hamermesh (1992).

2.   The most notable exception applies to authorship of graduate theses and dissertations. The student's major professor is not listed as a coauthor, although it is widely recognized that he or she probably played a major substantive role. That role also is the reason for the major professor typically being a coauthor of publications from the graduate student's work.

IV

Appendices

VI

Appendices

Appendix A

Example of a Research Proposal for a Government Agency

This proposal was submitted to the National Research Initiative Competitive Grants Program, U.S. Department of Agriculture, in 1995. It was selected for funding. Darren Hudson, a Ph.D. student at the time, now on the faculty at Mississippi State University, assisted in preparation of the proposal. Only the main body of the proposal is presented; numerous required forms and other specifics that vary among agencies (endorsement letters, biographical resources, and budget) have been omitted.

Project Proposal

Submitted to

United States Department of Agriculture

National Research Initiative Competitive Grants Program

Title: Cotton Price Policy and Trade Analysis

Principal Investigator: Don Ethridge

Texas Tech University

Key Words: Price controls, international trade, policy,

welfare economics, cotton, textiles

PROJECT SUMMARY

Internal policies of a group of cotton-producing and textile-exporting countries affect production, consumption, and trade patterns of cotton and textiles among the countries and in global markets. Specifically, taxation of the raw fiber sector through effective price ceilings and export taxes have been used to subsidize domestic textile production and exports in countries that practice these policies. More understanding is needed about effects of these policies on domestic supply, demand, and welfare in the home countries, as well as the effects of these variables on international trade and welfare distributions in the rest of the world, including the United States. The general objective of this project is to produce estimates of those effects.

To accomplish the objective, domestic supply and demand elasticities and export demand will be estimated and effects of the policy scenarios simulated. Assistance on literature searches for existing elasticity estimates and the data for elasticity estimation are being provided by the International Cotton Advisory Committee (ICAC) and the World Bank. A mathematical simulation approach, based either on structural econometric models or mathematical simulation using prior elasticity estimates, will be used to estimate changes in production, consumption, and welfare from the cotton price policies.

The research has the potential to be important to long-range U.S. policy in two primary ways: (1) It will foster understanding of the structure and implications of foreign cotton price policies on international trade and welfare; and (2) Estimates of the impacts of GATT can be derived from this analysis. The researchers on this proposed project have a history of expertise in cotton and textiles and access to a range of technical expertise on both cotton and textiles.

PROJECT DESCRIPTION

INTRODUCTION

Countries have long used commodity and trade policies to affect national goals. Industrialized nations have gravitated to more liberalized trade policies through such instruments as the General Agreement on Tariffs and Trade (GATT) and the North American Free Trade Agreement (NAFTA). However, trade restrictions such as nontariff trade barriers (NTBs) and internal policies are still a pervasive fact (Salvatore, 1993). The governments of nations also intervene in their domestic economies and trade sectors. The perceived needs to gain market share, secure foreign exchange, stimulate economic growth, and enhance industries' positions in the world economy have led governments of some nations to attempt to manipulate their markets to the advantage of their domestic industries.

Intervention in markets by governments can redistribute wealth by adjusting prices and/or quantities produced. Government intervention is generally thought to have two motives. The first is to protect domestic industries from competition. Government intervention in this case leads to economic rents accruing to those industries that are protected, which leads to socially wasteful rent-seeking (Brander and Spencer, 1985). This rent-seeking has been shown to lead to the operation of the nation inside its transformation curve (Krueger, 1974).

With respect to developing countries, a second reason for government intervention is to secure foreign exchange and to stimulate economic growth (Bhagwati and Srinivisan,

1979). These types of countries produce primarily raw products. They then export these products (in raw or semiprocessed forms) in exchange for manufactured goods (Anderson, 1992). Most nations in the global economy practice some type of intervention in their domestic commodity markets, the potential effects of which are numerous. Some implications are obvious, while others are harder to discern.

Policies in Cotton

The world cotton economy consists of 100–200 million growers, and a similar number of people employed seasonally in the production and harvest of 19 million tons of lint per year (ICAC, 1992). In 1992–93, 77% of the world cotton production was concentrated in China, the United States, India, Pakistan, and Central Asia. At the same time, 58% of world consumption was concentrated in China, the United States, India, Pakistan, and Brazil. Thus, the policies of these countries toward their domestic cotton industries affect global welfare and a significant portion of world production and consumption.

Townsend and Guitchounts (1994) divide government intervention in the raw cotton fiber sector into three primary categories: (1) extensive state control; (2) managed domestic prices; and (3) free market conditions. In both the extensive state control and managed domestic price categories, the prices for raw fiber are usually held below "world prices," either to secure trading profits for government marketing boards or to subsidize domestic textile mills or both. These "antifarmer" price policies, used primarily in developing nations (Gardner, 1990), tax the production of raw cotton. Policies in these countries may also include export taxes and quotas, import taxes and quotas, input subsidies, acreage controls, and production credits.

Pakistan and India, for example, fall into the category of managed domestic prices (Townsend and Guitchounts, 1994). These countries manage their cotton sectors so as to ensure relatively stable domestic prices for their producers and lower fiber prices for their textile mills, while allowing excess stocks to be exported at international market prices. Prices paid to producers in these countries are usually lower than free market prices, which allows domestic spinners to purchase cotton fiber at prices below the international level, thus conferring a competitive advantage since the cost of acquiring cotton accounts for between 49% and 61% of the total cost of yarn production (ICAC, 1992).

These policies have led to increases in cotton yarn production. For example, Pakistan has increased cotton yarn production by 46% over the 1986–90 period (Asian Development Bank, 1991). In 1991, cotton yarn exports from Pakistan represented 31% of the world trade in cotton yarn (ICAC, 1992). This is directly attributable to raw cotton prices in Pakistan averaging 24% less than border prices (Asian Development Bank, 1991; Hamid et al., 1990). An earlier study found a 21% implicit tax on producers of cotton in Pakistan for the 1981–87 period (Ender, 1990). Between 1988 and 1991, the average difference between internal and international market prices for raw cotton fiber was 20.6 US ¢/lb. (ICAC, 1992). There were also export taxes levied on the export of cotton yarn. However, this tax only represented 2.4 US ¢/lb., or about one-tenth of the subsidy discussed above.

Gains made in the production of cotton textiles have been offset to some degree by losses in the cotton production sector. Cotton production increased only 17% over the 1986–90 period, and Pakistan's raw fiber exports have almost ceased to exist. In fact, Pakistan will likely become a net importer of raw cotton fiber in the near future (Townsend, 1994). Countries with similar policies to Pakistan (e.g., India, Egypt, and Syria) (Evans, 1978; Townsend and Guitchounts, 1994; Townsend, 1994) have experienced similar responses in cotton fiber and yarn production. However, very little analysis exists on the effects of these policies on production, consumption, trade, and welfare.

Ender (1990) analyzed the effects of government intervention in Pakistan using producer subsidy equivalents (PSEs) and consumer subsidy equivalents (CSEs). He found that there was an average total implicit tax on cotton fiber producers of 12.4% of the producer price over the 1981–87 period. Conversely, there was an average subsidy to cotton spinners of 5,065 rupees/ton, which represented 30.2% of the spinners' price for raw fiber.

Fundamental economic theory shows that altering the operation of the market by manipulating price or supply alters the allocation of resources. Changing these relationships changes welfare for consumers and producers, and sometimes causes net social losses from the policy changes. In a market such as cotton, any policy that forces changes at one level in the marketing chain will filter welfare changes through the entire marketing channel (Kaiser et al., 1988). There is also evidence that suggests that policies such as those discussed have implications on the international markets through trade mechanisms (McCalla and Josling, 1985). Understanding the effects of these policies on trade and welfare is important. The new GATT regulations also will impact on the countries with policies as discussed, and on all cotton producers, consumers, and traders.

Varangis and Thigpen (1995) estimate that the most substantial impact of the GATT treaty will come in the cotton industry (i.e., raw fiber, textile, fabric, garments, etc.). The GATT treaty will begin phasing out trade restrictions on these items. Estimating the effects of the internal policies helps understand how GATT will change global production and trade flows.

STUDY OBJECTIVES

The general objective of the proposed study is to determine how developing countries' cotton policies that cap producer prices for the purpose of subsidizing the domestic textile sector affect production, consumption, trade, and welfare distributions in those countries. The specific objectives are to:

1.   identify countries with price policies that constrain cotton prices below world market prices in order to foster development of the domestic textile industry;
2.   determine changes in the countries' production, consumption, and trade in both the raw fiber and cotton textile markets;
3.   determine welfare effects of the cotton policies in both the raw fiber and textile markets for both the home country and the rest of the world; and
4.   estimate the effects of the GATT treaty on the production, consumption, trade, and welfare of the countries under study.

RATIONALE AND SIGNIFICANCE

The relevant economic sector in Indistan (a hypothetical country representing the policies described above) can be divided into two subsectors: (1) the raw cotton fiber market and

(2) the cotton yarn market. The raw fiber market represents the production and sale of raw cotton fiber to either domestic yarn spinners or exporters. The cotton yarn market represents the production of cotton yarn for either export or domestic fabric mill consumption.

Assume initially that both the raw fiber and yarn markets are structurally competitive, both internally and internationally (Martin, 1937; Cheng, 1988), and ignore transportation costs and exchange rate effects. The following sections examine expected effects of policies on the cotton fiber market and the cotton textile market.

Cotton Fiber Market

Figure 1 illustrates the cotton fiber market, where the panels represent Indistan, rest of the world (ROW), and the trade market. It is assumed that Indistan is a large-country exporter; Indistan is a sufficiently large producer that it can influence the world price for cotton (Gardner, 1990; Helmberger, 1991), but does not have monopoly power because its cotton has close, but not perfect, substitutes. Sirhan and Johnson (1971) state that every country that exports cotton is to some degree a price maker.

In figure 1, the intersection of the excess supply (XS) and excess demand (XD) functions yields a world price (Pw) and a quantity traded (X) before the implementation of any policy. This results in a quantity produced in Indistan of Qe and domestic spinner consumption of Qd. In the ROW, the total quantity demanded at Pw is equal to Qi. Producers in the ROW will supply Qp, while the remainder will be imported.

Indistan implements a "two-price" policy on raw cotton fiber. There is a Minimum Export Price (MEP), which is the minimum price at which a base grade of cotton can be exported from Indistan. This price is a subjective estimation of the world price that is set by a government committee (in practice, the MEP in Pakistan, for example, is related to the Cotlook B Index and the New York Futures price [ICAC, 1992; Townsend, 1994]). Thus, it

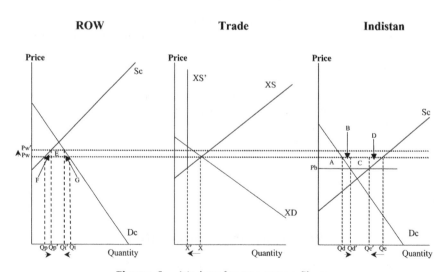

Figure 1. Markets for raw cotton fibers.

is assumed for this discussion that the MEP is analogous to the "world price." There is also a benchmark price, which is the internal price of cotton FOB port of export. This price does not represent a price given to producers. Rather, it is the price used by the government of Indistan to calculate the export tax on raw fiber. The tax is equal to the difference between the benchmark price and the MEP. Thus, the tax is equal to the world price minus the benchmark price.

There are two possible scenarios considered here for the two-price system discussed above. First, the benchmark price can be set above the domestic equilibrium price. Thus, exporters can purchase cotton in Indistan for less than the benchmark and sell at the world price. Since the exporter is purchasing raw fiber for less than the benchmark, the revenue from resale in the export market is sufficient to cover the tax imposed by the government, and exports are possible.

Figure 1 shows the effects of the two-price policy when the benchmark is set above the domestic equilibrium. The excess supply function would continue along its original path up to the point where the benchmark price intersects the domestic supply function. At that point, the tax on exports would be larger than the revenue the exporter could obtain, making exports no longer feasible. This would cause the excess supply function to become vertical at that point (Xs'). This will cause a decrease in exports (X to X'), and an increase in the world price (Pw to Pw'). This causes a decrease in total cotton production in Indistan (Qe to Qe') and an increase in cotton production in the ROW (Qp to Qp').

These changes cause redistributions that can also be seen in figure 1. Since the benchmark price in Indistan lowers the effective price of raw cotton fiber within Indistan, there is an increase in domestic consumer (yarn spinner) surplus. This can be seen by area A. Conversely, the benchmark price lowers the effective price to producers (versus what could be obtained in a free market at the world price), which causes a decrease in producer surplus of area (A + B + C + D). The government gains area C as revenue from the export tax. The remainder, (B + C), represents a net social loss to Indistan from the two-price policy. Since domestic producers would have been willing to produce more, and domestic consumers would have been willing to buy less at the original world price, (B + D) is transferred to the ROW. The export tax (effective price ceiling) raised the world price, which caused the ROW to produce more and reduce imports. Therefore, the rents from the two-price policy, (F + G) in figure 1, accrue to the producers of the ROW. However, consumers in the ROW lose area E. Thus, the net effect of the policy on the ROW is (F + G − E), the size of which depends on the relative magnitudes of the areas (elasticities of supply and demand).

The second scenario is for the benchmark price to be set below the internal equilibrium price. In this case, the revenue that an exporter can obtain from trading cotton would not cover the tax on exports. Thus, exports are not feasible and the entire cotton crop is reserved for the domestic spinning industry. This scenario is likely when the government of Indistan feels that domestic spinners will require the whole crop for their production needs.

A second portion of Indistan's policies affecting the raw fiber industry is an export quota. In the context of this analysis, the export quota is of secondary importance because the two-price system dictates how much cotton is exported. The export quota only distributes the licenses to export that quantity. Thus, the rents from the restricted export supply and the quota accrue to the exporters who hold licenses. There may be additional economic losses due to rent-seeking (Deacon and Sonstelie, 1989; Krueger, 1974), but these are beyond the scope of this analysis. Therefore, the export quota is ignored here.

Cotton Yarn Market

Policies affecting the raw fiber sector have direct effect on the yarn-spinning (textile) sector; these two sectors are highly related in terms of costs and revenues (supply and demand functions). The two sectors are presented separately only for graphical clarity.

The cotton yarn market is depicted in figure 2. Producers in this case are yarn spinners and the consumers are textile mills that produce cotton fabrics, garments, and so on, or exporters. Without a two-price system in the cotton fiber sector, the world price of cotton yarn is Pw and exports are X. Spinners in Indistan produce Qe, while the ROW produces Qp. If a two-price policy is implemented in the raw fiber market in Indistan, the policy lowers the cost of raw fiber to the Indistan spinning industry and shifts the supply function for cotton yarn (St to St'). This also shifts the excess supply function (XS to XS'), ceteris paribus. Under the large-country assumption (Gardner, 1990; Helmberger, 1991), this would decrease the world price of cotton yarn from Pw to Pw', and result in an increase in exports from X to X'. In Indistan, the quantity of cotton yarn produced increases and domestic consumption of yarn also increases. In the ROW, the quantity demanded increases, but production decreases.

The producer surplus that accrues to cotton spinners is (C + D − A − B). As Just and Hueth (1979) point out, in a multimarket framework in which one industry produces the major raw product (cotton fiber) for another industry's production process (cotton yarn), the consumer surplus in the lower market may be equal to the producer surplus in the higher market under some assumptions. Thus, the increase in producer surplus (C + D − A − B) in figure 2 may be equal to the consumer surplus in figure 1. However, this is subject to the primary assumption of monotonical price transformation between market levels, which may or may not hold in this situation. Despite this, the consumer surplus of cotton yarn in the domestic market has increased by area A (figure 2).

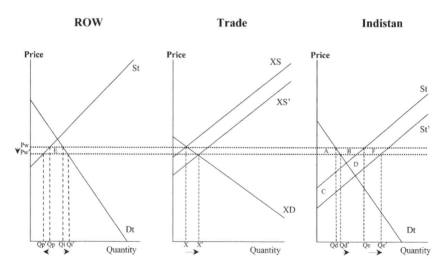

Figure 2. Markets for cotton yarn; no export controls.

Area B represents a net social loss in the cotton yarn market. This amount is transferred to consumers in the ROW in the form of lower prices for cotton yarn. This displaces some of the production by yarn spinners in the ROW. However, figure 2 indicates that decreases in producer surplus are offset by increases in consumer surplus. This may not always hold. The empirical result depends on the elasticities of supply and demand.

To capture some of the rents accruing to the foreign consumer, assume that the government of Indistan has also implemented an export tax on cotton yarn. Although this tax only slightly reduces the subsidy given to cotton yarn spinners, it does have the effect of capturing some of the transfers to the ROW. Figure 3 depicts the same situation as in figure 2 except that the initial shift in the supply function has been exaggerated so that the welfare effects of the export tax are easier to discern.

The export tax shifts the excess supply function from XS' to XS". Since the tax is assumed to only partially offset the subsidy, the shift will not take the excess supply function back to its original level of XS. This change has the effect of raising the world price of cotton yarn to Pw" and decreasing exports from X' to X". If it is assumed that the export tax is equal to Pw" − Pw', there will be no change in the quantity produced in Indistan. However, the quantity demanded in the ROW increases to Qi", and the production by the ROW increases to Qd". The net effect is to offset some of the changes from figure 2.

The government of Indistan collects area (B + C) as a tax on the quantity Qe' − Qd" (figure 3). This represents part of area B in figure 2. However, area (A + D) still represents a net social loss to Indistan that is transferred to the ROW. This gain is represented by area E in figure 3.

Thus, the net expected effect of the two-price policy in the raw fiber market is to decrease fiber production and exports, transfer wealth from the producers of cotton to cotton spinners in the form of a subsidy, and to transfer wealth from Indistan to foreign producers of raw fiber. In the market for cotton yarn, the two-price cotton fiber policy

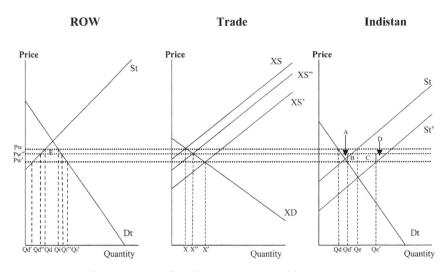

Figure 3. Markets for cotton yarn; with export tax.

stimulates production of cotton yarn. The magnitude of this change is affected by the price elasticity of demand for raw fiber and by the productive efficiency of spinners in handling the larger volume of cotton fiber. The increased production of cotton yarn lowers the world price of yarn, the extent depending on the elasticity of export demand of the ROW and the elasticity of export supply in Indistan.

Potential Implications for the United States   The existence of the price-capping policies are not a limited-case scenario. Although different in the specific makeup, approximately 60% of world cotton production for the 1988–93 period came from countries that exhibit extensive control over internal raw fiber prices (Townsend and Guitchounts, 1994). The primary countries are China, countries of Central Asia, Egypt, Syria, India, Pakistan, and Zimbabwe (Townsend and Guitchounts, 1994; Ender, 1990; Evans, 1978). Thus, this policy affects a large portion of the market in which the U.S. raw fiber producers and textile mills must compete.

The United States produces an average 19.7% of the world's raw cotton fiber (ICAC, 1994). Thus, any changes in welfare distributions in the cotton fiber sector resulting from the two-price policies in developing countries impacts the cotton producers in the United States. Although the United States is not a major importer of raw fiber, it does import and export cotton textiles. Changes in world price change the patterns of trade in world markets. Additionally, given the decreases in cotton exports and increases in cotton textile exports from countries such as Pakistan (Hamid et al., 1990; Townsend, 1994), the United States and other countries are affected.

The proposed study is organized so that additional scenarios can be analyzed. That is, trade effects coming from the implementation of GATT can be examined by removing the trade restrictions covered by GATT and recalculating the results. This may also be important to U.S. policy makers in future decisions about policies.

RESEARCH METHODS AND PROCEDURES

Methods for achieving the four specific research objectives are presented here, recognizing that some details must be addressed as the research progresses. The set of countries with similar price policies toward the raw cotton industry (objective 1) will be identified through collaboration or consultation with the International Cotton Advisory Committee (ICAC), the Foreign Agricultural Service (FAS), and the World Bank. To date, primary countries appear to include Pakistan, India, and China. Other countries, such as Zimbabwe and Syria, also practice these policies. At the current time, the feasibility of including China is being considered. However, China as a centrally planned economy poses problems with the reliability of data and modeling of economic behavior. Thus, inclusion of China is still under consideration.

To determine specific policy tools used by the defined set of countries, existing literature, contacts with the ICAC, FAS, and within other international organizations such as the World Bank and countries being analyzed will be used. The comparison of policies itself may be useful to those in the United States involved with international trade and policy.

The second objective, to evaluate changes in production, consumption, and trade, will be based largely on elasticity estimates. An extensive search of existing literature will be done to locate existing elasticity estimates. When that is complete, the reliability of those estimates will be evaluated. Depending on the availability and reliability or previously estimated elasticities, two basic approaches will be evaluated. The first will be to simply utilize the existing elasticity estimates in the simulation of the policy effects. A second possibility for finding the elasticity estimates will be to econometrically determine the elasticities needed through structural models. Discussions with the ICAC and other international organizations indicate that assistance in locating existing elasticities will be available.

Once the relevant elasticities (i.e., domestic supply and demand, export supply, and export demand) have been determined for both the raw fiber and cotton yarn markets, estimated changes in production, consumption, and trade will be derived through mathematical simulation. Simulations will be based either on the econometric structural model or on mathematical (nonstochastic) simulation, such as linear elasticity models (Gardner, 1990). Simulations will extrapolate from the existing situation to a free market situation, showing changes in the relevant variables. Results from the mathematical simulations will also be used to evaluate the welfare changes in both the raw fiber and textile markets.

Finally, the simulation can be used to estimate the effects of GATT. This can be accomplished by removing those trade restrictions that are covered by GATT and then reestimating the simulation. Changes in production, consumption, trade, and welfare can then be estimated.

The explicit hypotheses are that there will be decreases in the export of raw cotton fiber from the countries under study, and increases in the production and export of cotton yarn from the two-price policies in the raw fiber sector. A net social loss of the countries' cotton fiber industry practicing this policy is expected. These social losses are expected to be transferred to the ROW. A net social loss is also expected in the textile industry. Finally, the effects of GATT are expected to be an increase in cotton yarn trade as well as cotton trade.

Data   A literature search will be initiated with the aid of the ICAC and the World Bank to locate existing elasticity estimated and data from which structural econometric models can be estimated. Additionally, FAS information and literature will be used. The plan is to evaluate effects of these policies by individual country. However, if data limitations prevent analyzing individual countries, an alternative may be to aggregate the data from all countries together into a pooled time-series, cross-national data set.

Dissemination of Results   Results of the proposed study would be provided through the use of scholarly journals and appropriate technical bulletins. This should allow researchers of varied interests access to results and methods.

Limitations   A limiting factor to the proposed procedures is that it fails to account for other implications such as black markets (Devarajan et al., 1989), rents accruing from queuing (i.e., losses from cotton not being processed efficiently because mills cannot handle increased volume [Deacon and Sonstelie, 1989]), or social losses coming from rent-seeking behavior (Krueger, 1974). However, these effects are expected to be minor in comparison to production, consumption, and trade effects.

Tentative Schedule   The first year will be spent in the data collection, literature search, and model formulation phases of the project. This phase will be centered around consultation with potential data providers and will consume considerable time in getting the data in a usable form. This will also be the phase where time will be dedicated to the compilation of policy mechanisms of the various countries to be studied. Finally, structural econometric models used to estimate elasticities will be formulated in this stage.

The second year (or second phase) of the proposed study will involve structural model estimation of elasticity estimates, mathematical simulation, and interpretation of results. During this phase, the majority of time will be spent in data analysis, results interpretation, and preparation of results for dissemination.

CREDENTIALS FOR THE PROPOSED RESEARCH

The researchers have an established history in research in various components of cotton and textile economics. The PI has conducted research on all segments of the cotton industry, including textiles. The researchers also have access to a broad range of technical expertise in both cotton and textiles; they are affiliated with The International Textile Center (ITC) and the International Center for Arid and Semi-Arid Land Studies (ICASALS) at Texas Tech, and they have working relationships with staff at The International Cotton Advisory Committee (ICAC), The World Bank, and Foreign Agricultural Services, (FAS), USDA, which allows access to data and literature not always found in university library systems. Additionally, their academic department has alumni who hold positions in some of the governments listed in this proposal.

REFERENCES

Anderson, K. "The Changing Role of Fibers: Textiles and Clothing as Economies Grow." *New Silk Roads*, K. Anderson, editor. New York: Cambridge University Press, 1992, pp. 2–6.

Asian Development Bank. *Asian Development Outlook: 1991*. Manila, Philippines, 1991, pp. 162–163.

Bhagwati, J., and T. Srinivisan. "Trade Policy and Development." *International Economic Policy: Theory and Evidence*, R. Dornbusch and J. Frenkel, editors. Baltimore, MD: The Johns Hopkins University Press, 1979, pp. 1–35.

Brander, J., and B. Spencer. "Export Subsidies and International Market Share Rivalry." *Journal of International Economics.* 18 (1985): 83–100.

Cheng, L. "Assisting Domestic Industries under International Oligopoly: The Relevance and the Nature of Competition to Optimal Policies." *The American Economic Review.* 78 (1988): 746–758.

Deacon, R., and J. Sonstelie, "Price Controls and Rent Seeking Behavior in Developing Countries." *World Development.* 17 (1989): 1945–1954.

Devarajan, S., C. Jones, and M. Romer. "Markets under Price Control in Partial and General Equilibrium." *World Development.* 17 (1989): 1981–1993.

Ender, G. "Government Intervention in Pakistan's Cotton Sector." Wash., DC: U.S. Dept. of Agriculture, Economic Research Service, Agriculture and Trade Analysis Division, Staff Report No. 9041, 1990, pp. 1–41.

Evans, R. "Cotton in Syria." Wash., DC: U.S. Dept. of Agriculture, Foreign Agricultural Service, Report No. FAS-M-280, Apr. 1978, pp 1–23.

Gardner, B. *The Economics of Agricultural Policies*. New York: McGraw-Hill Publishing Co., 1990, pp. 37–324.

Hamid, N., I. Nabi, and A. Nasim. "Trade, Exchange Rate, and Agricultural Pricing Policies in Pakistan." World Bank Comparative Studies. Wash., DC: The World Bank, 1990.

Helmberger, P. *Economic Analysis of Farm Programs*. New York: McGraw-Hill Publishing Co., 1991, pp. 1–75.

ICAC. "Background Information for ICAC Discussions of the Impact of Internal Policies in Pakistan on International Cotton Prices and the Spinning Industries of Cotton Importing Countries." Wash., DC: International Cotton Advisory Committee, Standing Committee, Attachment N to SC-N-391, 1992, pp. 1–5.

ICAC. "Cotton: Review of the World Situation." Wash., DC: International Cotton Advisory Committee, Standing Committee, Sept.–Oct. 1994, pp. 1–27.

Just, R. and D. Hueth. "Welfare Measures in a Multimarket Framework." *The American Economic Review*. 69 (1979): 947–954.

Kaiser, H., D. Streeter, and D. Liu. "Welfare Comparisons of U.S. Dairy Policies with and without Mandatory Supply Control." *American Journal of Agricultural Economics*. 70 (1988): 848–858.

Krueger, A. "The Political Economy of the Rent-Seeking Society." *The American Economic Review*. 64 (1974): 291–303.

Martin, R. *International Raw Commodity Price Control*. New York: National Industrial Conference Board, 1937, pp. 8–12.

McCalla, A., and T. Josling. *Agricultural Policies and World Markets*. New York: Macmillan Publishing Co., 1985, pp. 36–59.

Salvatore, D. "Protectionism and World Welfare: Introduction." *Protectionism and World Welfare*, D. Salvatore, editor. New York: Cambridge University Press, 1993, pp. 1–5.

Sirhan, G., and P. Johnson. "A Market-Share Approach to Foreign Demand for U.S. Cotton." *American Journal of Agricultural Economics*. 53 (1971): 593–599.

Townsend, T. Statistician, International Cotton Advisory Committee, Wash., DC. Personal communication, June 1994.

Townsend, T., and A. Guitchounts. "A Survey of Income and Price Support Programs." *1994 Beltwide Cotton Conferences, Proceedings*. Cotton Economics and Marketing Conference. Memphis, TN: National Cotton Council, 1994, pp. 401–405.

Varangis, P., and E. Thigpen. "The Impact of the Uruguay Round Agreement on Cotton, Textiles, and Clothing." *1995 Beltwide Cotton Conferences, Proceedings*. Cotton Economics and Marketing Conference. Memphis, TN: National Cotton Council, 1995.

Appendix B

Example of a Master's Thesis Proposal

This proposal illustrates a proposal for an M.A. thesis submitted to a graduate committee. In this case the problem statement makes use of data to establish the extent or severity of the problem/issue (HIV/AIDS). The conceptual framework, while basic or elementary, focuses on the research problem, yet is directly and obviously connected to the analytical model structure proposed in the methods and procedures section. The literature review is organized around components related to the AIDS problem rather than as a study-by-study summary of relevant literature.

THE MACROECONOMIC IMPACT OF A GENERALIZED AIDS EPIDEMIC IN THE RUSSIAN FEDERATION

M.A. Thesis Proposal

Shombi Sharp

Department of Economics

University of Colorado at Denver

May 13, 2002

Thesis committee:

Dr. Thomas Rutherford

Dr. Charles Becker

Dr. Barry Poulson

I.   GENERAL PROBLEM

The HIV/AIDS epidemic has resulted in over 36 million people infected and 20 million dead globally over the past two decades. It is estimated that AIDS now kills more people worldwide than any other infectious disease (Oxfam, 2002). While AIDS has been called "the most serious threat to social and economic progress in Africa today," and new epidemics are emerging with exponential speed in eastern and central Europe, in India and China, and throughout Southeast Asia, Latin America, and the Caribbean (Oxfam, 2002).

The collapse of the Soviet Union created a mixture of economic and social dislocation more severe than the Great Depression in the West (*N.Y. Times*, 2000). This resulted in what is now being called the "Russian mortality crisis" of 1990–1995, with the most precipitous decline in national life expectancy ever recorded in the absence of war, repression, famine or major disease. It has been estimated that Russia experienced approximately 1.6 million premature deaths in the first half of the 1990s (Bennet et al., 1998).

The burgeoning HIV/AIDS epidemic exacerbating already declining population growth, life expectancy, and health indicators in Russia compounds the country's socioeconomic problems. The World Bank estimates that AIDS and tuberculosis combined could cost Russia 1% of its GDP annually within four years (ICG, 2001).

Although official figures place the number of HIV-positive Russians at 130,000, more new HIV infections were reported in Russia during 2000 than all previous years put together (Bloomberg, 2000). The United Nations Joint Programme for HIV/AIDS (UNAIDS) national representative calls the country's AIDS growth rate the world's highest (BBC News, 2000), with over 1 million infections expected by 2002 (Bloomberg, 2000). A recent collaboration between Russian and American researchers in Orel Province (southwest of Moscow) found that HIV infections there had risen almost 34-fold from 1997 to 2000 (CNN.com). According to UNAIDS guidelines on classifying AIDS, a significant rise in the incidence of reported TB cases indicates transition from a stage 2 to stage 3 epidemic (Barnett and Whiteside, 2000). While there are six stages of increasing severity overall, UNAIDS estimates that even the most critically affected countries remain classified in stages 4 and 5. Evidence is mounting that the rapid spread of the AIDS virus in Russia has set the stage for an epidemic of tuberculosis (CNN.com).

The eastern European and Russian context is significantly different from that of Sub-Saharan Africa (SSA) and Latin America and the Caribbean (LAC). Intravenous drug use, not heterosexual contact, constitutes the primary vector for transmission. Still, Médecins Sans Frontières estimates there are 3 to 4 million drug users at risk (Bloomberg, 2000), with the majority of new infections occurring in young adults between the ages of eighteen and twenty-five (BBC News, 2000). Further, given sexual contact between drug users and nonusers and an increasing spread among sex workers (Bloomberg, 2000), an assumption that the epidemic will not escape its early-stage concentration is tenuous. The growing prevalence of other Sexually Transmitted Infections (STIs), between 200,000 and 400,000 annually for Russia, raises the possibility that a generalized AIDS epidemic may "bootstrap off of or run concurrently with the injecting drug epidemic" (*N.Y. Times*, 2001). This suggests that the low sero-prevalence levels (as opposed to growth rates) relative to SSA and LAC conditions should be viewed as an opportunity for immediate prevention efforts rather than complacency. Further, official rates may capture only "the tip of the iceberg" of actual infections.

II.   SPECIFIC PROBLEM

The linkage between AIDS and the macroeconomy is complex. In addition to overwhelming the health resources of developing countries, AIDS impacts the social, economic, and demographic underpinnings of development. Because it is often sexually transmitted, AIDS disproportionately impacts the sexually active and economically productive segments of a population. AIDS is characterized by a "slow onset" of eight to ten years. Therefore, a twenty-five-year old contracting HIV will not, on average, develop full-blown AIDS until the mid-30s.

A brief outline of key mechanisms by which AIDS impacts macroeconomic performance follows:

   • *Reduction in labor force numbers*, targeted at productive age segments, reducing overall labor productivity;
   • *Reduction in the number of workers and savers* relative to the total population (i.e., increased dependency ratios);
   • *Increase in wages and costs* resulting from labor force reduction, which in turn can lead to loss of international competitiveness;
   • *Lower public revenues and reduced national savings* (private and public) resulting from decreased labor force levels and unit productivity, lowering capital accumulation and slowing formal-sector employment creation;
   • *Increase public health care expenditures*, reducing public savings;
   • *Increase social safety net expenditures*, reducing public savings.

The World Bank country director for Russia points out that AIDS in Russia is still in an early stage (BBC News, 2000). Concentrated epidemics such as Russia's are more sensitive to early and focused prevention campaigns. The return on investment in working with groups such as IDUs and sex workers and their clients could exceed that on conventional capital investment (*N.Y. Times*, 2001). Further, due to the delayed onset of AIDS, even current behavioral intervention will have little impact on death rates for many years to come.

Critical questions remain concerning the nature and scale of the epidemic in Russia. Within the context of constrained budgets, the costs of inactivity must be weighed against those of policy interventions if public resources are to be allocated efficiently. The motivation for this study is the incomplete information in this complex policy issue where today's decisions may impact long-run macroeconomic potential.

III.   OBJECTIVES

The general objective of this study is to project the impact of the AIDS epidemic on the Russian economy. This will be achieved through the following specific objectives:

   • Project the demographic impact of a hypothetically generalized HIV/AIDS epidemic on the Russian population;
   • Identify key linkages between relevant demographic and economic variables within the Russian context;
   • Define impact of the HIV/AIDS epidemic as the difference between GDP growth paths of "no AIDS" and "with AIDS" scenarios of varying degrees.

IV. LITERATURE REVIEW

The economic consequences of demographic change have long been the focus of academic inquiry, sometimes with surprising results. To illustrate, the fourteenth-century bubonic plague, bringing a dramatic rise in mortality rates, caused the populations of European countries to decrease by one-third within fifteen years, but resulted in substantially increased real wages (Over, 1992). This suggests that model specification of the stylized epidemic in question plays a crucial role in determining whether or not a population and health-related shock increases or decreases economic activity on a per capita level.

The estimation of the macroeconomic "cost of illness" was pioneered by Rice (1966), using a dichotomization of the costs of disease into the "direct" and "indirect" average costs, which are then multiplied by the annual number of cases to arrive at a total annual cost of any one illness. Direct costs capture the public and private resources used for meeting patients' health needs, including preventive measures. Modeling direct costs is, at times, referred to as the "health sector perspective" (Bishai et al., 2000, p. 3).

Indirect costs are often calculated using a human capital approach, calculating the present value of the stricken individual's future earnings discounted to the time of death or initial illness (Cuddington, 1993). This is related to the "societal perspective" on the cost of illness, in which the loss of patients' and their families' productivity are captured (Bishai et al., 2000, p. 4). While these estimated approaches have limited application to the specific study of AIDS due to the lack of marginal and intertemporal effects, a number of authors have found it useful in related investigations. The human capital approach has been used recently to calculate the impact of the "Mortality Crisis," from 1.8% to 4.7% of GDP in the Russian economy (Bloom and Malaney, 1998).

Several researchers have investigated the macroeconomic effects of AIDS over the last ten years, focusing both narrowly on health care expenditures and more broadly on impacts on national savings, labor force size, and productivity. These attempts have covered a range of approaches/methods, producing differing results across approaches and countries. Studies in the early 1990s found that, as has been suggested of the bubonic plague, macroeconomic impacts tended to be relatively modest in terms of per capita GDP. Due to a number of factors, including the unexpected severity of unfolding epidemics, improved modeling techniques and greater appreciation of the breadth of linkages, renewed interest in modeling the macroeconomic impact of AIDS has emerged in the last two years (Arndt and Lewis, 2000). Cuddington (1993) used a Solow-style, single-sector growth model with two factors to simulate the impact of AIDS on aggregate and per capita growth in the Tanzanian economy. On the other end of the spectrum, Arndt and Lewis (2000) employed a fourteen-sector computable general equilibrium (CGE) model of the South African economy that includes a Total Factor Productivity effect of the epidemic. Both approaches address the marginal dynamics absent in the cost of illness and human capital approaches discussed above. Relevant to the present study in the Russian social context, Cuddington (1993, p. 175) notes that relative impact of the negative savings effect on public and private saving will depend to a large extent on the nature of the health care system. Over (1992) used two neoclassical Cobb-Douglas production functions to differentiate impacts between urban and rural sectors while maintaining the assumption of increasing marginal costs to AIDS infections.

Of particular concern in modeling the impact of AIDS in lower-income countries is the frequent presence of surplus labor in the economy. Theory suggests that countries with high unemployment in the formal sector tend to exhibit a labor supply elasticity such that a mortality-related reduction in labor supply has little or no effect on output (Bloom and Malaney, 1998, p. 2078). This might even translate into increased per capita income as workers involved in low-productivity activities fill the vacancies created by AIDS in the more productive formal sector (Cuddington and Hancock, 1994, p. 1). Therefore, models that do not account for surplus labor dynamics may well exaggerate the impact of declining labor supply.

There are, however, several reasons to believe this effect may not prevail. First, the macroeconomic drag on the economy through alternate channels may be sufficient to independently reduce the number of formal-sector jobs and, therefore, demand for labor. And, as Cuddington and Hancock (1994, p. 2) point out, the rigidity of the Russian labor market may pose additional constraints to rapid labor market adjustments:

> *In a sticky-wage economy, the inability of firms to distinguish HIV-infected persons or to lower wages to compensate for the reduced productivity due to AIDS will cause firms to hire less labor and switch to more capital-intensive production methods. This may further exacerbate the inefficient use of labor that typifies dualistic economies.*

To test these competing arguments, Cuddington and Hancock (1994) constructed a dualistic macroeconomic model for Malawi, where surplus labor prevails in the informal sector and formal-sector wages are "sticky" (in excess of market clearance price), slowing the speed of labor market adjustment. The authors then compared the labor-surplus model results with those from a full-employment version. Noting that the most striking aspect is the similarity of the results, the authors conclude that the impact of AIDS on capacity utilization, not the presence of underutilized capacity per se, is the relevant consideration. Cuddington (1993) applied the same method to Tanzania and again found impact to be roughly the same across surplus-labor and full-employment scenarios.

The structure of the economy in question is also important. Of particular relevance to Russia, studies in lower-income countries have found that "economies based on extractive industries or export agriculture are likely to be most severely affected" (BBC News, 2000, p. 6). In this regard, CGE models benefit from the added dimension of sectoral structure and impact analysis.

While not exhaustive, table 1 summarizes the models examined in the course of this literature search. Results vary considerably according to assumptions and the stage of the epidemic at the time the paper was written (references are included in the bibliography).

Several points related to prior studies may be noted. Due to limitations in knowledge about factors driving the global HIV/AIDS epidemic, the epidemiological projections underpinning these economic models have systematically undershot prevalent growth paths. Further, little analysis has focused on the macroeconomic impact of HIV/AIDS in Russia, likely due to the fact that AIDS has only recently been identified as a threat in Russia and the epidemic is in the early stages. The absence of sufficient data and/or the lack of access to them may have been a deterrent to such studies. Finally, all analyses have assumed constant savings/investment rates across scenarios, often with no foreign investment. This structure fails to account for the negative relationship between market expectations of future productivity loss (due to public knowledge of the epidemic) and investment posited in the theory of rational expectations.

Table 1. Summary of macroeconomic impact studies of AIDS

Year	Author(s)	Country	Model type	Time horizon	Adult HIV prevalence (%)	Impact on GDP growth (%)	. . . and per capita (%)*
1992	Cuddington and Hancock	Malawi	Solow	1985–2010	0.1 to 3.2	−0.2 to −1.5	−0.1 to B0.3
1992	Over	30 countries (SSA)	Solow	1990–2025	0.1 to 3.2	−0.56 to B1.47	+0.17 to B0.33
1993	Cuddington	Tanzania	Solow/ dualistic	1980–2010	0.1 to 3.2	−0.5 to B1.3	+0.1 to B0.7
1994	Cuddington and Hancock	Malawi	Solow/ dualistic	1985–2010	0.01 to 1.1	−0.4 (med. scenario)	0 (med. scenario)
2000	Nicholls et. al.	Trinidad & Tobago	CGE	1997–2005	0.9 to 7	−4.2	Not reported
2000	Nicholls et.al.	Jamaica	CGE	1997–2005	0.9 to 7.4	−6.4	Not reported
2000	Arndt and Lewis	South Africa	CGE w/TFP effects	1998–2010	6.7 deaths/100 workers	−2.6 growth, −17 to B22 level	−8 to B13 level

*Impact on annual growth, unless otherwise noted.

V.   CONCEPTUAL FRAMEWORK

The diagram in figure 1 illustrates the conceptual macroeconomic linkages that are hypothesized to be important in understanding how AIDS impacts the Russian economic system. Given the demographic structure of the country, population and health circumstances combine with the economic/industrial infrastructure and social policy via workforce levels, labor efficiency, and fiscal expenditures to determine national income and growth.

In a system in which there is no HIV/AIDS, there is some level of macroeconomic activity, determined by existing demographics and economic structure (which determines impact channels). The introduction of HIV/AIDS into the system affects population and health. These impacts, in turn, affect workforce productivity and the health care system, which then impact employment, output, income, and so on. Conceptually, the macroeconomic impact of the incidence of AIDS is the difference in GDP level and growth rate between the two situations.

VI.   METHODS AND PROCEDURES

VI.1   Model

This research will seek to model the impact of a generalized HIV/AIDS epidemic on the growth path of Russia, calculated as the difference between a baseline "without AIDS" and several "with AIDS" scenarios of varying degrees. This will be achieved using a Computable General Equilibrium (CGE) simulation model developed using the Generalized Algebraic Modeling System (GAMS) language. The model will feature feedback mechanisms among key economic blocks, including the neoclassical growth assumptions capturing the assumption that the marginal cost of each additional lost worker is higher than that of the last (Over, 1992). Population impact will be disaggregated according to productivity class as identified in previous AIDS impact studies. The model structure is shown in figure 2.

A baseline "no-AIDS" demographic projection over the modeling horizon will be constructed and subjected to a "with-AIDS" impact analysis based on a stipulated generalized epidemic. Labor efficiency units and the fiscal burden of AIDS-related health and social insurance costs constitute the explicit channels of impact flowing from the

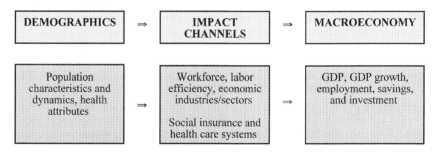

Figure 1.   Macroeconomic linkages important to an AIDS epidemic in Russia.

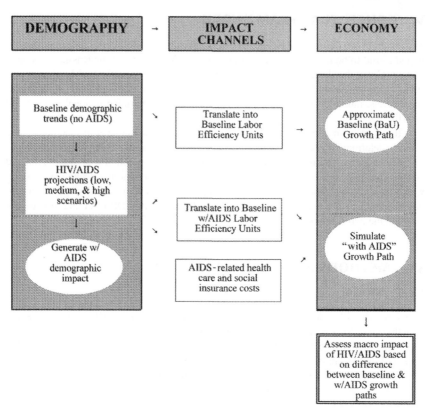

Figure 2.   Structure of the AIDS impact model.

demographic to the economic spheres. Both the baseline and with-AIDS scenarios will be run through this process to capture varying levels of specificity and assumptions. The results of each will be compared, with divergences in growth paths and other economic aspects constituting the impact of a stipulated HIV/AIDS epidemic. The approach will focus on the variegated effect of AIDS on population segments exhibiting different productivity levels, AIDS-induced on-the-job productivity loss, and the concurrent shift in the long-run equilibrium growth path due to changes in productivity, population, consumption, savings, and investment (i.e., capital formation). An attempt will also be made to account for the negative relationship that may exist between economic agents' expectations of future productivity loss, due to public knowledge of the epidemic, and investment. This hypothesis, largely unexplored in AIDS modeling, is based on the theory of rational expectations (Lucas, 1972).

The key advantage of the CGE approach is that, unlike partial equilibrium or aggregated Solow-style growth models, projections are based on actual microeconomic data. Output can therefore be customized to the nuanced economic conditions across sectors with sector-level impact readily observed. The Social Accounting Matrix (SAM) feature of

CGE models provides "key behavioral and accounting constraints . . . which in turn serves as an important check on the 'reasonability' of outcomes" (Arndt and Lewis, 2000). Finally, CGE models can be disaggregated to investigate the relative effects of changing structural parameters with potential policy implications. In this sense, the CGE model represents a simulation laboratory setting.

VI.2   Data

A drawback of the CGE approach is the procedure's need for a large amount of very specific input/output data across a sufficient number of productive sectors to arrive at plausible results. Access to periodically updated data sets, such as the GTAP series, has made this process manageable for the study of many countries. This study will use a modified version of the Rutherford-Paltsev multisector Russian economy data set (1999). AIDS-specific data to substantiate model parameters will also be collected.

VI.3   Epidemiological/Demographic Projections

This study will seek to identify suitable demographic projections on the AIDS epidemic in Russia for use in the CGE model. If prepared data are unavailable, original projections will be generated using existing methods. This would likely center on a framework combining the modeling methods of Schmitz and Castillo-Chavez (1994) and Bos and Bulatao (1992) along the lines of S. Nicholls et al. (2000). Such an exercise would require obtaining existing information on adult sexual behavior in Russia.

VII.   REFERENCES

Arndt, C., and J. D. Lewis, "The Macro Implications of HIV/AIDS in South Africa: A Preliminary Assessment," *The South African Journal of Economics*, 68 (no. 5, 2000): 856–887.

Barnett, Tony, and Alan Whiteside, *Guidelines for Studies of the Social and Economic Impact of HIV/AIDS*, Best Practice Collection, UNAIDS, Geneva, 2000.

BBC News, "Russia's AIDS Catastrophe," Nov. 16, 2000.

Bennet, Neil G., David E. Bloom, and Sergei F. Ivanov, "Demographic Implications of the Russian Mortality Crisis," *World Development*, 26 (no. 11, 1998): 1921–1937.

Bishai, David, Maria K. Lin, and C. W. B. Kiyonga, "Algorithms," *Bank Policy Research Working Paper 2321*, 2000.

Bloom, David E., and Pia N. Malaney, "Macroeconomic Consequences of the Russian Mortality Crisis," *World Development*, 26 (no. 11, 1998): 2073–2085.

Bloomberg. "World Bank Says AIDS, TB May Cost Russia 1% of GDP," April 23, 2000.

CNN.com, "Russia Facing AIDS Catastrophe," Nov. 16, 2000.

Cohen. "The Economic Impact of the HIV Epidemic," *UNDP HIV and Development Issue Paper* no.2.

Cuddington, John T., "Modeling the Macroeconomic Effects of AIDS, with an Application to Tanzania," *The World Bank Economic Review*, 7 (no. 2, 1993): 173–189.

Cuddington, John T., and John D. Hancock, "The Macroeconomic Impact of AIDS in Malawi: A Dualistic, Labour Surplus Economy," *Journal of African Economies*, 4 (no. 1, 1994): 1–28.

ICG, "HIV/AIDS as a Security Issue," ICG Report, June 19, 2001.

Lucas, R. E., "Expectations and the Neutrality of Money," *Journal of Economic Theory*, 4 (1972).

New York Times, "The Global Plague of AIDS," April 23, 2000.

New York Times, "Russia: Faces of a TB Epidemic," Nov. 1, 2001.

Nicholls, S., et al., "Modelling the Macroeconomic Impact of HIV/AIDS in the English Speaking Caribbean," *South African Journal of Economics* (2000).

Over, Mead, "The Macroeconomic Impact of AIDS in Sub-Saharan Africa," *World Bank Review* (June 1992).

Oxfam, "Development: The Import of HIV/AIDS," Oxfam Community Aid Abroad Web site, March 2002.

Rice, D. P., "Estimating the Cost of Illness," *Health Economics Series*, No. 6. DHEW Pub. No. (PHS) 947-6. Rockville, MD: U.S. Department of Health, Education, and Welfare, 1966.

Rutherford, Thomas, and Sergey Paltsev, "From an Input-Output Table to a General Equilibrium Model: Assessing the Excess Burden of Indirect Taxes in Russia" (Aug. 1999).

Appendix C

Example of a Ph.D.
Dissertation Proposal

This illustrates a graduate student proposal submitted to a graduate committee whose members were all economists. Note that in this case the specific problem statement contains extensive details on the structure, conduct, and performance of the pharmaceutical industry, deemed needed for the committee members to understand the main dimensions of the problem and justifying the objectives. Dr. Macy, now on the faculty at West Texas A&M University, completed her dissertation as proposed and has since published journal articles from that dissertation.

Dissertation Proposal

THE IMPACT OF PHARMACEUTICAL MERGERS

ON ECONOMIC AGENTS

Anne Marie Macy

Department of Economics and Geography
Texas Tech University

October 1998

Dissertation Committee:
Eleanor T. von Ende, Chairperson
Klaus Becker
Tom Steinmeier

General Issue:

Mergers and acquisitions are methods of corporate growth. When internal growth is not efficient or desirable, a firm may expand externally by acquiring another firm. Basic finance teaches that the main goal of a firm is the maximization of stockholder wealth. Therefore, the objective of an acquisition is to increase wealth. This goal may be achieved by increasing sales, market share, assets, or research, along with decreasing costs. Many firms and managers equate size with power. Mergers allow for an increase in critical mass, which the firm and managers believe will give them increased power in the marketplace. Mergers are a part of the business strategy of a firm and help create a sustainable competitive advantage for the buyer firm through the addition of the positive and desirable attributes of the acquired firm.

A merger is the combination of two firms, with the stockholders of both firms jointly owning the new firm. A merger forms a new entity. An acquisition occurs when one firm purchases the assets of another firm with the acquired shareholders ceasing to be owners. The acquired firm becomes a part of the buyer firm. A takeover is an acquisition where the buyer firm is substantially larger than the acquired firm is. The definitional differences are important for accounting and regulatory purposes but have little economic or financial meaning. Therefore, the terms *merger, acquisition*, and *takeover* are used interchangeably for purposes of this study.

Mergers in the twentieth century have come in waves. Houlihan et al. (1997) denote six main periods of merger activity. From 1893 to 1904, merger activity was dominated by industry consolidation leading to monopolistic entities. This was partly due to the Sherman Antitrust Act of 1890 that outlawed collusion but not mergers. Between 1915 and 1929 the second wave of horizontal mergers formed oligopolies. Mergers in the 1940s were caused by heavy estate taxes that forced small owners to sell as they retired. Most of these mergers were friendly. The theory of diversification propelled the longest merger wave, lasting from the mid-1950s through 1969. Medium-sized firms acquired other medium-sized firms in unrelated industries, forming conglomerates. A firm conducting business in several unrelated firms was believed to be less susceptible to the ill effects of business cycles. This wave ended when the stock market dropped in the early 1970s. Starting in the mid-1970s, firms began divesting unrelated or unprofitable divisions, the opposite of a decade earlier. The 1980s saw a continuing of the disassembly of conglomerates. The speed of takeover activity increased with the use of junk bonds. As the 1980s ended, there was a slowdown in merger activity due to the stock market drop and instability of financing from the use of junk bonds. The sixth wave started after the stock market drop and increased with the rise in the stock market in the 1990s. Strategic mergers have characterized this wave. Firms merge or acquire in order to gain specific products, markets, or skills. As shown in table 1, 5,848 mergers, net of cancellations, were announced in 1996. This figure is the highest in twenty-seven years but below the 6,107 announcements in 1969. The percentage of merger deals with a value over 100 million dollars has also increased during the last decade, indicating that the merging firms were already large.

Mergers and acquisitions are part of the market for corporate control. The aim is the right to manage the assets of the target firm. Successful management will lead to an increase in shareholder wealth. However, principal-agent problems may exist between management and the stockholders. Managers may undertake mergers to increase their importance and make other decisions that maximize their utility instead of stockholder

Table 1. Historical figures on U.S. mergers

Year	Net M&A announcements	Total dollar value paid ($ billion)	1992 constant dollar value ($ billion)	Number of $100 million deals
1968	4,462	43.6	157.4	na
1969	6,107	23.7	81.7	na
1970	5,152	16.4	53.6	na
1971	4,608	12.6	39.3	na
1972	4,801	16.7	49.9	15
1973	4,040	16.7	47.2	28
1974	2,861	12.4	32.2	15
1975	2,297	11.8	28.0	14
1976	2,276	20.0	44.8	39
1977	2,224	21.9	46.1	41
1978	2,106	34.2	67.2	80
1979	2,128	43.5	78.7	83
1980	1,889	44.3	73.3	94
1981	2,395	82.6	125.0	113
1982	2,346	53.8	76.6	116
1983	2,533	73.1	99.9	138
1984	2,543	122.2	161.0	200
1985	3,001	179.6	228.5	270
1986	3,336	173.1	214.8	346
1987	2,032	163.7	197.0	301
1988	2,258	246.9	286.8	369
1989	2,366	221.1	246.5	328
1990	2,074	108.2	115.6	181
1991	1,877	71.2	73.2	150
1992	2,574	96.7	96.7	200
1993	2,663	176.4	171.9	242
1994	2,997	226.7	215.9	383
1995	3,510	356.0	330.9	462
1996	5,848	492.9	446.9	640

Sources: Houlihan et al., 1997; GDP deflator: *Economic Report of the President*, February 1996.

wealth. The decision to merge is optimally based on expected value created by the merger and the sources of the value creation. Efficient decision making regarding initiation of mergers demands full information about the target firm.

The actual integration of the firms determines success. The initial decision is usually based on optimistic evaluations of the target firm in terms of its financial, strategic, and organizational strengths. The weaknesses of the target firm and the difficulties of actual integration are often overlooked or underplayed. Determinism, value destruction, and leadership vacuum are three sources of integration problems (Sudarsanam, 1995). *Determinism* refers to the conflict between the managers of the two merging firms. The acquirer does not account for the uniqueness of the target firm and forces rigid rules and programs on the new employees. If the managers and employees from the target firm believe that they are being overlooked and ignored, they may act in ways to prohibit or delay the integration, demonstrating their determinism. If the two sets of managers and employees are not able to function together, the value of the new entity will be reduced from the inefficiency. Friction between the two parties can cause communication lapses, which reduce the productivity of the entire firm. A third problem is that a clear leadership hierarchy is not always known, and the top management may not be involved in the tedious aspects of integration. This lack of leadership further leads to integration problems and a loss in wealth. Clearly, the three integration problems are variations of the principal-agent problem. A successful merger must account for these potential problems, which can delay the gains from the acquisition, and for the role employees play in the actual merger outcome.

Specific Problem:

In the pharmaceutical industry, there were twenty-one major mergers between 1989 and 1994. While this was during a period of somewhat reduced merger activity for the United States as a whole, pharmaceutical firms were combining with each other, insurance providers, biotechnology, and over-the-counter firms for strategic purposes. A maintained hypothesis is that the merger activity springs from the environment: that mergers were a response to the changing environment faced by the pharmaceutical firms. The environment has two main components, internal and external factors.

The first step in analyzing a business action is to examine the environment the firm faces. Pharmaceutical firms maneuver among various stakeholders. The firms' first responsibilities are to the shareholders, who are interested in returns on their investments. The returns are generated from profits, which are created from the difference between sales to customers (total revenue) and costs to produce (total costs). Sales are determined by consumer choice, which are influenced by physicians, insurance companies, and pharmacists. Firms can attempt to persuade choice through advertising and by providing an array of products.

At the center of the interplay among the players is the research and development (R&D) process. Success in the pharmaceutical industry is driven by a firm's ability to innovate. This process is long and expensive, demanding large cash flows to support it. While the winner of the process receives a government patent, which can assist in controlling the marketplace (and hence, revenues for the firm), the patent does not guarantee profitability.

Society's desires for more drugs and cheaper drugs are, to some extent, mutually exclusive. More and better drug innovations cost money and time. Consumers rarely are willing to pay for the entire discovery process, which takes about twelve years from start to finish. The final cost is high because many compounds are unsuccessful. Not even the added incentive of a patent is by itself enough to induce the risky behavior. While a patent gives exclusive marketing rights to the holder, it does not mean that there are no competitors. A patent is on the chemical composition of the drug, not the therapeutic use. Therefore, several drugs, all under patents, can be competing in the same area. Grabowski and Vernon (1990) examined one hundred new drug innovations from the 1970s. They found that only the top thirty drugs covered the average R&D costs, with the bottom ten of this group only barely covering costs. Thus, the cost to innovate a drug is not necessarily offset by the barrier-to-entry properties of a patent.

Proprietary drugs manufactured by the firms face competition from generics after the patents expire. The prevalence of generic substitution and the ease with which a generic can receive FDA approval after the Waxman-Hatch Act has increased the market share of generic drugs. In 1984 generics controlled 18.6% of the domestic drug market. By 1990 it was 32.9% and 41.6% in 1996 (IMS America, 1997). Typically, the generic drugs also quickly gain market share. Within one year of generic entry, generic drugs control an average of 61% of the market. The expected rapid loss of market share reduces predictions about the proprietary drug's overall return. This puts further pressure on cash flow and research and development. A new drug innovation costs around $350 million, whereas a generic drug costs only $1 million to bring to market (PhRMA, 1997a).

In addition to prescription drugs, many pharmaceutical firms also develop and market over-the-counter (OTC) drugs. Patients choose and self-medicate themselves with OTC drugs, many times without any professional input. Therefore, for a drug to be allowed to be marketed as an OTC, it must be extremely safe and the probability of an overdose or a severe side effect must be very small. OTC drugs do not provide wide margins for the drugstore or manufacturer, but their sales tend to be steady and positively correlated to advertising. Hence, many firms now use OTC drugs to provide a stream of cash to fund R&D. Bristol-Myers Squibb efficiently uses OTC drugs and cosmetics such as Bufferin, Comtrex, and Miss Clairol to finance its innovation process.

Not only does the R&D process affect firms by providing products and profits, it is a drain on cash flow. Firms manage their flows in order to fund all viable projects in one time period, which will fund the next projects in future time periods. Cash flow represents what the firm can spend on R&D, other internal projects, or external distributions. The need for a steady cash flow encumbers the firms to produce drugs to feed the cash flow. One of the external distributions is interest payments to bondholders. If cash flows are decreased through the R&D process, the ability to make the interest payments could become questionable. Any threat to liquidity threatens bond valuation and bond ratings. Bondholders want successful R&D but not at their expense.

Overall, U.S. firms conduct 36% of world pharmaceutical research and development. Japan accounts for 19%, followed by Germany (10%), France (9%), the United Kingdom (7%), and Switzerland (5%). The total dollar figure of pharmaceutical R&D in the United States was $11 billion in 1994 and is expected to be over $19 billion in 1998. Conversely, the entire research budget of the National Institutes of Health (NIH) for 1997 was $13 billion (PhRMA, 1997a). Pharmaceutical research represents only a part of the NIH's research program. As shown in table 2, domestic R&D expenditures have grown faster than expenditures abroad in almost every year since 1981. The growth in domestic expenditures has been steadier than foreign expenditures. This is partly due to the R&D process. The quality and quantity of research technology and personnel induce firms to conduct their research in the United States. Also, the FDA requires that the majority of clinical testing occur in the United States.

Consumers rarely want drugs to treat ailments. The demand for pharmaceuticals is derived, first, from the demand for health and second, from the demand for health services. The demand is further complicated by the various agency relationships of physicians, pharmacists, and managed care organizations that are elements of the distribution chain. The demand is also influenced by diagnostic categories. Respiratory and circulatory system diseases are associated with higher drug use than gynecological and nervous system conditions (Schweitzer, 1997).

Table 2. Growth in research and development, U.S. and abroad, 1970–94

Year	Domestic R&D—U.S. and foreign firms ($ million)	Annual change (%)	R&D abroad—U.S. firms only ($ million)	Annual change (%)	Annual total ($ million)	Change (%)
1970	566.2	—	52.3	—	618.5	—
1971	626.7	10.7	57.1	9.2	683.8	10.6
1972	654.8	4.5	71.3	24.9	726.1	6.2
1973	708.1	8.1	116.9	64.0	825.0	13.6
1974	793.1	12.0	147.7	26.3	940.8	14.0
1975	903.5	13.9	158.0	7.0	1,061.5	12.8
1976	983.4	8.8	180.3	14.1	1,163.7	9.6
1977	1,063.0	8.1	213.1	18.2	1,276.1	9.7
1978	1,166.1	9.7	237.9	11.6	1,404.0	10.0
1979	1,327.4	13.8	299.4	25.9	1,626.8	15.9
1980	1,549.2	16.7	427.5	42.8	1,976.7	21.5
1981	1,870.4	20.7	469.1	9.7	2,339.5	18.4
1982	2,268.7	21.3	505.0	7.7	2,773.7	18.6
1983	2,671.3	17.7	546.3	8.2	3,217.6	16.0
1984	2,982.4	11.6	596.4	9.2	3,578.8	11.2
1985	3,378.7	13.3	698.9	17.2	4,077.6	13.9
1986	3,875.0	14.7	865.1	23.8	4,740.1	16.2
1987	4,504.1	16.2	998.1	15.4	5,502.2	16.1
1988	5,233.9	16.2	1,303.6	30.6	6,537.5	18.8
1989	6,021.4	15.0	1,308.6	0.4	7,330.0	12.1
1990	6,802.9	13.0	1,617.4	23.6	8,420.3	14.9
1991	7,928.6	16.5	1,776.8	9.9	9,705.4	15.3
1992	9,312.1	17.4	2,155.8	21.3	11,467.9	18.2
1993	10,477.1	12.5	2,262.9	5.0	12,740.0	11.1
1994	11,101.6	6.0	2,347.8	3.8	13,449.4	5.6

Source: PhRMA, 1997a.

Drug innovations are classified as *breakthroughs* and *me-too's*. A breakthrough is an entirely new way to treat an illness, a new therapeutic class within a therapeutic market. Me-too's are drugs for which other drugs in a therapeutic market already exist. While consumers gain from me-too drugs, which provide greater selection and possibly lower prices, the breakthroughs represent gains in treatment quality.

The therapeutic areas that have highest demand and potential revenues are those areas in which the firms specialize. In order to encourage more innovations for lesser diseases, Congress passed the Orphan Drug Act of 1983. Orphan drugs are drugs that treat diseases that affect fewer than 200,000 people. Because the potential markets are so small, pharmaceutical companies may not have an incentive to innovate for these markets. Under the act, innovating firms are eligible for tax credits to cover part of the development process and are given exclusive marketing rights for seven years. From 1984 to 1993, one in three new drugs approved were considered orphan drugs.

Compounding the complicated R&D process was the push for lower prices. Managed care organizations (MCOs) and the federal government wanted more drugs because drug therapies are less expensive than most invasive procedures, but demanded cost control. Currently, the United States spends over 13% of gross domestic product (GDP) on health expenditures. This figure is double the 1970 percentage and is a higher percentage than the other major industrialized countries. However, domestic pharmaceutical expenditures constitute only 1.2% of GDP, which is equivalent to those in Canada and less than those in Italy, France, Germany, and Japan (PhRMA, 1997b). While pharmaceuticals constitute only 5% to 6% of health care expenditures, the industry is an easy target due to its size, availability, and recognizability. Physicians and hospitals are more disaggregated around the country and harder to differentiate. Part of the visibility of the pharmaceutical costs is because the out-of-pocket expenditures on pharmaceuticals is 39.5% of all outpatient drug expenditures, which is higher than the 2.4% of out-of-pocket expenditures on hospital services. Private health insurance pays 40% of all outpatient expenditures, while federal and state Medicaid pay over 17% (Health Care Financing Administration, 1995).

While domestic pharmaceutical expenditures appeared consistent with figures from other industrialized countries, the price inflation of pharmaceuticals has been rising faster than the overall rate of inflation. During the 1970s, the urban consumer price inflation rate averaged 7.1%, whereas the prescription drug component of the CPI was only 3.6% (table 3). In the 1980s, the overall rate was 5.6% and the drug component grew at 9.6%. From 1990 to 1997, all items averaged 3.3%, whereas the drug prices grew at 5.3%. Starting in 1981 prescription prices started to grow faster than the urban consumer rate, and it has in each year since then except 1985. The drug component of the producers' price index also exceeded the urban CPI for most of the 1990s.

The federal government has also passed laws and set policy concerning the pricing behavior of drugs. The laws have a great effect because the government is a large purchaser of pharmaceuticals through the Veterans' Administration and Medicare/Medicaid. The Omnibus Budget Reconciliation Act of 1990 began providing federal matching funds to state Medicaid for prescriptions covered under a manufacturer rebate agreement. The rebate is calculated as the difference between the average manufacturers' price and the lowest price paid by any buyer. There is also a minimum rebate of about 15% of the average manufacturing price. The Veterans' Health Care Act of 1992 began requiring that the drug manufacturers enter into pricing agreements with the Veterans' Administration, Department of Defense, and Federal Supply Schedule in order to do business with the federal government. Currently, the drugs purchased under this act cannot exceed 76% of the nonfederal average manufacturing price of the most recent year. Moreover, price increases are strictly limited by increases in the urban consumer price index. The 1996 total rebate was approximately $2 billion, up from $553 million in 1991 (PhRMA, 1997a). A partial result of this rebate is a shift of costs from the government to private insurers and individuals who do not receive rebates and, therefore, must pay the full amount for each drug.

The growth in managed care organizations and their emphasis on cost containment has led to a variety of techniques to control drug expenditures, which puts pressure on firms' revenues. Private third-party reimbursement of pharmaceuticals has increased from 29% in 1990 to 60% in 1996 (IMS America, 1996). The first technique is the use of formularies, lists of drugs that are approved for insurance coverage. Formularies can be open,

Table 3. Inflation rates, 1970–97

Year	CPI—all items (%)	CPI—drugs (%)
1970	5.7	1.7
1971	4.4	0.0
1972	3.2	−0.4
1973	6.2	−0.2
1974	11.0	2.3
1975	9.1	6.2
1976	5.8	5.3
1977	6.5	6.1
1978	7.6	7.7
1979	11.3	7.8
1980	13.5	9.2
1981	10.3	11.4
1982	6.2	11.6
1983	3.2	11.0
1984	4.3	9.6
1985	3.6	9.5
1986	1.9	8.6
1987	3.6	8.0
1988	4.1	8.0
1989	4.8	8.7
1990	5.4	10.0
1991	4.2	9.9
1992	3.0	7.5
1993	3.0	3.9
1994	2.6	3.4
1995	2.8	1.9
1996	3.0	3.4
1997	2.3	2.6
Average 70–79	7.1	3.6
Average 80–89	5.6	9.6
Average 90–97	3.3	5.3

Source: Bureau of Labor Statistics, various years.

in which coverage extends to all drugs with reduced copayments for the use of particular drugs, or they can be closed, which restricts coverage to only listed drugs. Drugs not on the formularies have lower revenues. Therapeutic substitution, the replacement of one drug for another, usually in the same therapeutic class, is another technique to contain expenditures. The use of substitution has increased from 22% in 1990 to 57% in 1995. HMOs also switch patients from a proprietary drug to its generic equivalent. Formularies and therapeutic substitution can reduce the profitability of the firms.

Because drugs are a basic component of health care, insurance companies, employers, other managed care organizations and Medicaid now use pharmacy benefit managers

(PBMs). The PBMs process and review the prescribing and use of drugs. Many PBMs also fill the prescriptions through large mail-order businesses. PBMs currently manage benefits for about 50% of all Americans who have pharmaceutical insurance coverage. Because of the importance of PBMs, the top three firms have been acquired by major pharmaceutical firms.

Some pharmaceutical firms reacted to this changing environment by merging. As indicated in table 4, three types of mergers occurred. Horizontal mergers were transacted between firms of similar size. Examples of this type are Bristol-Myers and Squibb, Glaxo and Wellcome, Upjohn and Pharmacia, Ciba-Geigy and Sandoz, and SmithKline and Beecham. These mergers combined R&D facilities and staffs, sales staff, geographic markets, and major products to increase critical mass and to facilitate competitiveness in domestic and international markets. Because R&D staffs work in teams, combining teams can enhance creativity. Research teams are also expensive to produce, making it advantageous for some firms to acquire the existing staff of another firm. Increasing sales staff generates wider exposure for all the drug products. Doctors tend to prescribe what they know. An increased marketing staff can better inform physicians, encouraging them to prescribe more. The time delay from the research and development process has caused a need for blockbusters or "wonder" drugs to be produced by a firm to sustain it over the long run. Horizontal mergers increase the likelihood of producing a blockbuster drug through the combined resources.

Alternatively, vertical mergers between pharmaceutical firms and insurance providers have been used to take advantage of the growing use of managed care organizations in health care delivery. Of particular interest were the PBMs that influence the choice of drugs through formularies and drug utilization boards. After the 1994 Eli Lilly–PCS merger, the Federal Trade Commission established a fire wall to separate the management decisions of owner pharmaceutical firms and their pharmacy benefit managers in order to control collusive behavior. However, even with the fire wall, an easing of communication between the entities may lead to a decrease in costs that may, in turn, encourage increased use of the drugs manufactured by the owner pharmaceutical firm. Other examples of this type of merger are Merck's acquisition of Medco and SmithKline Beecham's acquisition of Diversified PBM.

Finally, pharmaceutical firms have responded to the changing industrial environment by acquiring biotechnology firms and OTC firms. Typically, OTC firms provide a steady cash flow that may be used to fund the long R&D process. OTC products have a relatively steady demand and are cheaper to produce than prescription drugs because of less stringent regulations. The research potential of biotechnology has not yet been proven but is expected to pay off in the future by revolutionizing the way in which pharmaceuticals are developed. Traditionally, research has been conducted by looking for compounds and then determining if there are any clinical applications. Biotechnology research looks instead at the disease or ailment and tries to develop something to control or eradicate it. It involves knowledge of genes, cells and entire systems. The increase in computer-related technology is instrumental to this process. By attempting to develop something explicitly for an ailment, it is expected to increase success rates of innovated compounds, thereby decreasing costs and increasing possible blockbuster innovations. An example of this type of merger is American Home Products' acquisitions of A. H. Robins and American Cyanamid, with which they also received Immunex.

Mergers represent a strategic response to internal and external factors surrounding the pharmaceutical industry. The increase in size may provide an answer for the firms. The

Table 4. Major mergers in the pharmaceutical industry, 1989–97

Year	Bidder	Target	Type of Merger
1989	Dow	Marion	Horizontal
1989	Bristol-Myers	Squibb	Horizontal
1989	American Home Products	A. H. Robins	OTC
1989	American Cyanamid	Praxis	Biotechnology
1990	Pharmacia	Kobi	Horizontal
1990	Boot	Flint	Horizontal
1990	Abbott Laboratories	Damon Biotech	Biotechnology
1990	Baxter	BioResponse	Biotechnology
1991	SmithKline	Beecham	Horizontal
1991	American Home Products	Genetics Institute	Biotechnology
1991	Sandoz	Systemix	Biotechnology
1991	Boehringer Mannheim	Microgenics	Biotechnology
1993	American Cyanamid	Immunex	Biotechnology
1993	Merck	Medco	Vertical
1994	SmithKline	Sterling	OTC unit
1994	Sanofi	Sterling	Rx unit–Horizontal
1994	Pharmacia	Erbamont	Horizontal
1994	Hoffman-LaRoche	Syntex	Horizontal
1994	American Home Products	American Cyanamid	Horizontal
1994	Eli Lilly	PCS	Vertical
1994	SmithKline Beecham	Diversified	Vertical
1995	Glaxo	Wellcome	Horizontal
1995	Hoechst	Marion Merrell Dow	Horizontal
1995	Upjohn	Pharmacia	Horizontal
1995	Rhone-Poulenc Rorer	Fisons	Horizontal
1996	Ciba-Geigy	Sandoz	Horizontal
1997	Roche	Boehringer Mannheim	Horizontal

Source: *Mergers and Acquisitions*, various issues.

new, larger firms may be better able to produce drug innovations, to market their products, to deal with managed care organizations and the federal government, to increase cash flow, and to increase power in the market place. The goal of a merger is to gain control. For the pharmaceutical industry, the firms want to increase control not just of the attributes of another firm but also to increase the ability to exert control over their environment. Mergers are beneficial if they are wealth producing for the economic agents.

Mergers affect various groups associated with the mergers. While the mergers were undertaken as a result of changes in the environment, the firms were not the only stakeholders affected. Stockholders, to whom the managers of the firm are primarily responsible, are obvious stakeholders affected by the mergers. The firm has other stakeholders who are also concerned with the outcome of the merger. Bondholders can lose value if the firm undertakes additional debt or if the merger is unsuccessful. Consumers' interest in the merger lies with the number, quality, and price of new drugs innovated. If the merger results in more products, consumers gain. Losses are possible if the quantity or quality of production declines. The stakeholders are concerned not just with the immediate effects but also the longer-term effects of the mergers. Gains or losses to the stakeholders may not be realized immediately. A longer time frame allows for the inclusion of the integration process and its success or failure. While a successful business earns net income and increases in the stock price, longer-term success is interdependent with the valuation of the firm by the other stakeholders. Therefore, each group has a definition of a successful merger strategy.

Objectives:

The primary objective of this study is to determine the effects of three specific mergers (Bristol-Myers–Squibb, Merck-Medco, and American Home Products–American Cyanamid) within the pharmaceutical industry on industry stakeholders and their implication for the industry in general. The specific objectives are to determine the effects of the mergers on:

1. Bidder stockholders' returns;
2. Target stockholders' returns;
3. Firm operating performance (the effect on return on assets);
4. Other stakeholders—bondholders and consumers (effect on bond ratings and whether more drugs and what types of drugs were innovated).

Literature Review:

While extensive research has been conducted on the pharmaceutical industry, there are no prior studies of pharmaceutical mergers. The majority of the research on the pharmaceutical industry is centered on topics such as the profitability of the industry, the role of research and development, the role of generics, and the impact of institutional influences. Research on mergers is also extensive, with topics focusing on the motivations for mergers and empirical results of mergers. Therefore, to understand the role of mergers within the pharmaceutical industry, both sets of literature are examined. Pharmaceuticals are discussed in the first section, and mergers in the second. Each section is further divided into subsections on different elements of each part.

Pharmaceutical Industry Literature

The majority of the work on the pharmaceutical industry focuses on the role that research and development plays in the risk and returns of the firms. The first three articles discussed consider the costs a known fact. It is not until the Grabowski and Vernon (1990, 1993) articles and the DiMasi et al. (1991) article that quantitative measures of the current cost of R&D are studied. Other areas of consideration for this study are the role of firm size in innovations, the pricing behavior of the firms, and international considerations that may encourage mergers.

Risks and Returns to Research and Development   Comanor (1986) discussed the industry and presented a synthesis of prior work. He analyzed the profitability of the industry first in terms of market structure. He noted that while the industry may appear to be monopolistic, it is actually highly competitive. The firms compete not in price competition but product competition. New products, especially generics, are constantly entering the market, forcing the existing products into obsolescence. Comanor added that a second factor pressuring the firms is technology, which is also rapidly changing and expensive. He viewed the R&D process as a lottery with winners and losers. The cost of the technology to innovate is so expensive that Comanor recommended viewing R&D as a public good with financial support from the government. After assessing for the various factors facing the industry, Comanor concluded that the industry did not earn above-average profits for a manufacturing industry and may have earned profits well below what was needed to compensate the firms for the risk of the R&D process. In his discussion of technology, Comanor did not mention that increases in technology can lead to several discoveries instead of just one, each helping of offset the cost of the technology. He also ignored the role computer technology can play in the discovery process. His discussion lacked empirical evidence, which weakened his argument on the expensiveness of the industry and need for government assistance.

Scherer (1993) addressed the profitability of the industry during the Clinton health care debate. He presented the R&D process as the reason for the high pharmaceutical costs. Unlike Comanor, Scherer did not believe that generics were a concern to the firms and the R&D process. Instead, he considered the federal regulation to be a serious threat. In a similar conclusion to Comanor, he advocated less regulation and more assistance for the pharmaceutical firms. Scherer argued that the pricing pressures and lack of patent protection around the world faced by the industry would lead to fewer innovations. However, the fault for the decrease in production lies with the world governments, not the industry. The article was clearly written to assist in the formation of opinions at the time, but it lacked empirical evidence to support his assertions.

Also critical of federal regulation was a study by the Boston Consulting Group (1993). This article presented evidence on the increased regulatory presence of the Food and Drug Administration (FDA) on the conduct of the pharmaceutical industry. The FDA had increased the number of regulatory employees in the prior five-year period, and the number of pages of submissions to the FDA also increased during this time. The paper hypothesized that between the increased regulation and pricing pressures by managed care organizations, R&D expenditures would fall, as would the international position of the industry. The group recommended that firms consider working with or merging with the managed care organizations and improving relations with the FDA in order to maintain marketing power. While this paper was sponsored by Pfizer, an American pharmaceutical firm, which may cause the motivation for the study to be questioned, it does present the concerns of the firms at this time and addresses a potential role for mergers in the industry.

The returns to drug innovations were discussed by Grabowski and Vernon (1990). They found that of the new drugs innovated in the 1970s, only thirty of one hundred covered average R&D costs, with ten barely covering costs. The authors concluded that the R&D process was very risky and costly. The authors drew no policy implications for the firms. Grabowski and Vernon (1993) continued their research and examined the returns to drugs in the 1980s. They found that the internal rate of return was 11.1% with a mean net present value of $22 million. This figure is slightly less than the figure estimated by the Office of Technology Assessment (1993), which found a net present value of $36 million. Grabowski and Vernon accredited the differences to accounting techniques. Both studies concluded that the pharmaceutical industry was risky and needed government support and not regulation in order to assure continued innovations.

DiMasi et al. (1991) quantified the cost of R&D. Examining ninety-three innovations in the 1970s, the authors found that a new drug innovation cost over $200 million in the 1970s, and the figure had increased to over $300 million for drug innovations in the late 1980s. The authors systematically examined each stage of the research process and attributed a cost figure to each step. The amounts were then discounted at the cost of capital of 9%. One drawback of the results is that the authors make no distinction between development costs per therapeutic class. Also, since most firms specialize in a few therapeutic areas, their costs may not be as high if a base amount of knowledge already exists.

Firm Size   If innovation is the heart of the industry, a natural question is whether large or small firms produce more innovations. Acs and Audretsch (1988) examined innovation in firms of all sizes and industries and found that while large firms produced more innovations in total, smaller firms innovated more on a per-employee basis. Closer examination of their data found that the pharmaceutical industry was prolific and required more expensive labor to innovate than other industries like appliances and metals.

Greer (1992) examined the pharmaceutical industry and found that the smaller firms produced more innovations than the larger firms because the larger firms were mired in bureaucracy. Most of the firms Greer cites were biotechnology firms. A drawback of this study is that Greer did not account for the strategic alliances between many biotechnology firms and large manufacturers. It could have been that the large pharmaceutical houses contracted out their R&D in order to decrease risk and costs.

Pricing Behavior   The pricing behavior of the industry relates to mergers by indicating the extent of pricing power firms can exert in the marketplace. If firms have limited power, it may be advantageous to combine to increase market power, or if generics limit the longevity and profitability of drugs, an increase in size may be needed to produce more drugs to counter the influence of generics.

Caves, Whinston, and Hurwitz (1991) addressed the pricing behavior of the pharmaceutical industry during 1976–1987. They found that the average proprietary drug's price rose after patent expiration and then fell when generic products entered the market, but only by 4.5%. Generic products' prices tended to be 10% less than the proprietary drug's price but gained little market share. Grabowski and Vernon (1992) also examined pricing patterns of generics during the 1980s. They found that the entry of new generics was directly related to the brand-name profit margin. The uniqueness of the Grabowski and Vernon study is their finding that generic products exhibited first-mover advantages and pharmacists and physicians had quality perceptions about the generics. The generics competed against each other and not just against the proprietary drug. Neither sets of authors discussed the role that managed care organizations could have played in the usage and pricing behavior of generic drugs during the 1980s.

Lu and Comanor (1994) examined the level of price competition among new drug innovations during the 1980s. They found that drugs that entered the therapeutic markets where a drug already existed were priced lower to gain market share and then were increased in price over time to match the price of the original drug. The authors did not address if the drugs were profitable for the firm or if the price increases were an attempt to recoup costs.

International Considerations   Grabowski (1989) addressed the effects of pricing pressures on the amount of funds available for research and the international competitiveness of American pharmaceutical firms. He found that the domestic firms innovated more drugs and of higher therapeutic value than any other country. However, he warned this could end in the 1990s if generics and managed care organizations reduced profits. Schweitzer (1997) examined this and found that the European firms were strong competitors to American firms during the 1990s. He also found that countries with extensive price controls on pharmaceuticals had high total health expenditures because consumers substituted other health services. He indicated that domestic firms needed to compete more aggressively with their European counterparts or risk losing international power. Both authors mentioned but did not expand upon the idea that merging into larger firms was a way the domestic firms may be able to compete internationally.

Merger Literature

Relevant merger research is divided into two main areas: theoretical reasons for mergers and empirical studies of the effects of mergers on stockholder value. The articles on theoretical reasons are mostly presentations of ideas with no empirical context. No single explanation encompasses all merger activity. Instead, there exists a variety of motivations. The empirical articles use event studies to test the various theories and examine trends in merger activity. Most of the event studies examine price changes with a short time frame centered around the date of the merger announcement.

Theoretical Reasons for Mergers and Value Increases   Agency problems and efficiency gains are the two broad categories of merger theory. If agency problems motivate a merger, managers act in their own interest with the possibility of financial harm to the firm's stockholders. The total gains to a merger under this circumstance tend to be nonpositive. Mergers may also be undertaken for possible synergy gains. In this case, the merger gains will be nonnegative. The two categories are not competing theories but instead provide a framework in which to analyze a merger.

Jensen (1986) credited agency problems and the lack of free cash flow payouts as the cause for mergers. Managers are the agents of stockholders and are charged with acting in the best interests of those stockholders. Free cash flow is the cash available in excess of the cash needed to fund profitable projects. Payouts of free cash flow to stockholders reduce the funds available to managers, which reduces the power of managers because they have less cash under their control. Hence, managers are reluctant to distribute funds. A solution to this problem is to spend the free cash flow on acquisitions, which can be value-enhancing for the firm. Firms with large cash reserves are more likely, according to Jensen, to acquire other firms.

Schleifer and Vishny (1988) have expanded upon the agency conflict idea. They charged that the managers of acquiring firms do not always act in the best interest of their stockholders. Instead, mangers use the extra available cash flow to buy new projects. This behavior may reduce the value of the firm at the same time it satisfies the utility of managers, who now are part of a larger company or a company newly invested in an up-and-coming field. This causes returns to shift from the bidder stockholders to the bidder managers. The bidder firm may overpay for the target firm. This shifts gains from the bidder stockholders to the target stockholders.

The agency theory espoused by the prior two articles is problematic. If the agency theory offered a comprehensive explanation, one would expect more hostile takeovers than currently occur as firms try to dump excess cash. Also, if gains are always shifted to target stockholders, eventually bidder stockholders would realize this and no would want to be the bidder. While managers may not always make decisions in the best interest of stockholders, efficient market theory tells us that a manager who did so would receive funds from stockholders fed up with inefficient management. A lag may occur as stockholders determine efficient management, but eventually returns should go to the efficient firms and managers.

The winner's curse or hubris hypothesis is derived from the agency conflict hypothesis. Roll (1986) noted that the winner of a takeover bid is willing to pay more for the target firm than any other firm would pay. This is because the winning bidder overestimates the value of the target. In essence, the winner's hubris leads to the curse. Even without competition for the target, Roll suggests that bidders overestimate and overpay during the acquisition process. Capen, Clapp, and Campbell (1971) also found that winners are cursed by overpaying for acquisitions. A shortcoming of this theory is that its proponents have looked at a very narrow time frame, usually six months before the offer to six months after the offer, in their discussions of gains to stockholders. Modern finance theory dictates a longer-run view of mergers. It may take years to reap the benefits of a merger. While stock price fluctuations around the merger offer reflect future expectations about the merger, actual gains and synergies may not be realized for years.

Efficiency gains from a merger may occur from a signaling or information effect. Weston, Chung, and Siu (1998) discussed the idea that a takeover offer signals to the market that the bidder sees more value in the target than the market currently does. If the market agrees with the undervaluation, the target's stock price will rise. Bradley, Desai, and Kim's (1988) analysis indicated that the information from a takeover bid could encourage the target's management to undertake activities to increase efficiency regardless of the outcome of the takeover bid. Takeover offers can indicate a firm that is not living up to its potential. The gains in stock price go to the target shareholder for the market recognition of the firm's undervalued or inadequate performance. Share prices of targets that received unsuccessful bids did not fall back to preoffer levels for several years.

The q-ratio hypothesis relates the market value of a firm's securities to the replacement costs of the firm's assets (Weston, Chung, and Siu, 1998). When this ratio falls below one, it is cheaper for the firm to buy an operation as opposed to building its own. This is one possible cause for mergers in the 1970s. The q-ratio provides an explanation for mergers between firms conducting the same type of business within an industry. The q-ratio is more relevant for industries that rely on technology and equipment instead of human capital.

Fama and Jensen (1985) noted that mergers that diversify a firm allow a decrease in risk for undiversified stakeholders. It is questionable whether mergers for diversification are helpful for diversified stockholders, who can diversify on their own. Managers and employees may benefit, and the benefit may be transferred to stockholders. Theories of management teams imply that employees and managers who work together develop certain efficiencies and skills. If the team is divided, the efficiencies are lost. If the team is transferred to a new operation, the team skills remain intact. This benefit can result in lower costs and increased efficiencies, which can translate into higher returns to stockholders.

The main theories on merger value rely on synergy as the improvement mechanism. Synergy is called the "2 + 2 = 5 effect." The sum of two parts is greater than the two parts alone. Weston, Chung, and Siu (1998) noted that many horizontal mergers occur because it is believed that the two firms make a good fit and complement each other. The merger can compensate for deficiencies in each of the firms. Operating synergy may result in managerial economies from all parts of the firm, including production, research, marketing, and finance.

Operating synergy can also occur from vertical integration. Arrow (1975), Klein, Crawford, and Alchian (1978), and Williamson (1975) each discussed the value from a vertical merger. The combination of two firms at different stages of production or distribution of a product can decrease the cost of communication. The increase in coordination allows for a smoother function system, which potentially creates efficiencies and profits.

The synergy theories are rooted in the belief that the firms are not initially operating efficiently. If the firms are inefficient, then the merger represents a gain not only to stockholders but also to society. The concern over concentration is moot for mergers that are wealth increasing. Synergy theories also assume that economies of scale exist over all parts of the cost curve or at least for the area of concern. If this were true for all firms, then it would be desirable for all firms to condense into one.

None of the theories is fully able to explain all merger motivations. However, the fundamental ideas lay the groundwork for an examination of merger activity.

Empirical Tests of Merger Activity   Positive returns to targets are well documented in the literature. Jensen and Ruback (1983) reviewed thirteen studies of merger performance. Target stockholders averaged returns from 20% to 30%. Jarrell, Brickley, and Netter (1988) studied 663 tender offers from the early 1960s to the mid-1980s. They found that returns to targets ranged from 19% to 35%. The results from the Bradley, Desai, and Kim (1988) study mirrored the results of Jarrell, Brickley, and Netter.

While returns to target shareholders are positive, returns to bidder shareholders are not as clear. Jensen and Ruback (1983) found returns to bidder stockholders ranged from 0% to 4%. Jarrell, Brickley, and Netter (1988) and Bradley, Desai, and Kim (1988) obtained similar results. However, a weakness of these studies was the length of the time period used to generate these results. All studies used windows of less than two months to determine the returns. With such a small time frame, the studies are not necessarily looking at returns to stockholders from a merger but returns to stockholders from a merger offer. The returns from a merger may not appear or be known immediately.

The method of payment used in a merger may also influence the returns to stockholders (Myers and Majluf, 1984). The use of stock sales to pay for an acquisition implies that the stock is overvalued. Debt financing implies undervaluation of the stock. Wansley, Lane, and Yang (1983) studied mergers in the 1970s and found that target firms received a higher abnormal return when cash was used instead of stock. They did not look at debt financing. Travlos (1987) estimated returns for bidders and target stockholders. His results indicated that returns to target stockholders averaged about 12% for stock transactions and 17% for cash transactions. Bidder stockholders received negative returns in stock transactions and less than 1% in cash transactions. A major drawback to the Travlos research is that he used a two-day announcement period as the time frame for analysis; results based on a two-day announcement period are not dynamic enough to capture the full impact of mergers.

Berkovitch and Narayanan (1993) tested for total returns with different merger theories and found that the correlation between target returns and total returns was positive and significant for mergers that exhibited total positive returns. The correlation between positive target returns and positive bidder returns was not significant. The authors attributed this to the hubris theory. Concerning the mergers that resulted in negative total returns, the authors found that target gains were negatively correlated with the negative total returns and with the bidder returns. The authors pointed to the agency theory to explain the negative total returns and bidder returns, but believed that the hubris theory explained the positive gains to targets. Of the mergers examined, 76% exhibited positive total returns. The authors concluded that the agency and hubris theories may have accounted for the distribution of returns among stockholders, but the mergers occurred because of synergy.

Healy, Palepu, and Ruback (1992) analyzed the postmerger performance of fifty large mergers. They examined operating, investment, and employment variables. The authors found that employment decreased and operating measures increased. The investment variables did not change significantly. Agrawal, Jaffe, and Mandelker (1992) studied over a thousand mergers that occurred in the 1970s and 1980s. They found that acquiring-firm stockholders realized a 10% loss in wealth over a five-year period following the merger. Unlike in most other studies, the authors adjusted for size and risk due to the large variety of firms in the study.

Industry effects are another important consideration in the analysis of merger activity. Dess, Ireland, and Hitt (1990) noted the need to account from a management viewpoint for industry effects and a longer time period. Mitchell and Mulherin (1996) and Subramanian et al. (1992) each accounted for this effect. Results from Mitchell and Mulherin found that most mergers in the 1980s were in industries that had experienced noticeable shocks such as deregulation and changes in oil prices. They also discussed the role foreign competition and innovation have in interindustry patterns and concluded that typically the industries reacted to broad economic and financial shocks instead of industry-specific shocks.

Subramanian et al. analyzed the food and kindred products industry. They looked at firms from a homogeneous industry in order to examine mergers from a more microeconomic scale and to compensate for industry effects. The authors also used a wider time frame, five years before and after the merger announcements, primarily to look at returns to bidder stockholders who may not have received returns immediately. Results suggested positive returns to the stockholders and improvements in operating performance. While focusing on a specific industry over an expanded time frame has benefits, the empirical results of the study are limited because they are based on only ten years of annual data. The use of monthly or quarterly data would have dramatically raised the power of their empirical results. A drawback of the longer time frame is that the data may have been influenced by external factors not related to the merger (Lubatkin and Shrieves, 1986). Therefore, a trade-off exists between characteristics of data and the quality of the measure of performance.

Conceptual Framework:
Objective of Firms

A basic premise of economic analysis is that a firm seeks to maximize profits. Profits are defined as total revenue minus total costs. Total revenue is determined by sales and the sales price. Total cost is the product of average total cost and sales. Therefore, a firm seeks to make the difference between total revenue and total costs as large as possible.

However, static profit maximization does not necessarily provide the proper guidance to managers (Keown et al., 1998; Brigham and Gapenski, 1993; Moyer, McGuigan, and

Kretlow, 1990) because profit maximization does not include a time component. It provides no way to compare short- and long-term decisions and benefits. Managers must examine the trade-offs between short-run and long-run returns to determine in which profit-maximizing opportunities to invest.

A second shortcoming of static profit maximization is that no consistent definition of profit exists. Accountants can alter the counting of costs and revenues in the calculation of profit. This leads to hundreds of possible figures even if standard accounting rules are followed. Even if a clear definition of profit is determined, whether the goal of the firm should be to maximize total profit, the change in profit, or earnings per share is still undetermined.

Simplistic profit maximization also does not account for risk differences among projects. Two projects may provide identical expected cash flows, but the risk of the future flows may differ between the two projects. The risk can be from the timing of the project, the financial risk of the investment, or the business risk of the investment. A firm can increase the return from an investment by leveraging it; however, the increase in earnings per share causes an increase in financial risk. An efficient market will recognize that the leveraged project entails more risk and value it accordingly.

Because static profit maximization does not provide enough guidance, managers instead attempt to maximize the market value of stockholder wealth. Stockholder wealth is the market price of a firm's stock. The market price represents the discounted expected future returns from owning the stock, which can come from dividends or price changes. It is the market value and not book value of the stock with which investors are concerned. The book value is a historical concept and is not indicative of future earnings potential. The market value is the price at which the stock trades in the marketplace.

Three main attributes of a firm's cash flow determine the market value of a share of stock. The first factor is the amount of future cash flows expected to be generated for the benefit of stockholders. The greater the expected flows, the greater the value. The second factor is the timing of the cash flows. The third factor is the perceived risk of the cash flow. The longer the time needed to receive the cash flow and the more uncertain the cash flow, the higher the return must be to compensate the stockholders. In determining the value, the flows are discounted at an appropriate discount rate, which represents the opportunity cost of the investment. It takes into account the returns that are available from alternative investment opportunities during the specific time period.

Cash flow is the determining factor because it can be spent, whereas net income cannot be dispensed since it does not reflect the actual inflows and outflows of cash. Cash flow is concerned with the true cash generated by the firm that can be used to acquire assets, make expenditures, and be distributed to investors. Cash flow can be raised externally or internally. External cash flow comes from equity and debt. Internal cash flows come from operations and the sale of assets. Initially, the cash flows are used to acquire assets that are used in the production of products. The production should result in funds for reinvestment and for distribution to creditors and owners.

Capital markets act to control the system and enforce efficiency. A firm's stock price is its single comprehensive index of success. It reflects the investors' best evaluation of the firm's true earning power. The process is prospective; it discounts the firm's future prospects into the current market value of the stock. It is comprehensive by considering all corporations and rendering judgments about the quality of the current and expected performance in a widespread, rapid, continuing process of evaluation. The evaluations reflect choices—hence they are comparative, realistic, and comprehensive.

If a pharmaceutical firm's decision to merge was correctly based on efficiency gains and not agency problems, the merger should have a positive effect on cash flows. Initially, the integration process may require a cash outflow. However, once the synergies are realized, cash flows should increase. An efficient market will recognize the gains, and stock price returns should reflect the value-increasing merger. Cash flows may decrease due to agency problems or not be as large if the potential synergies were overestimated. Cash flows that are less than expected should cause stock price returns to decrease.

Stockholders are not the only people interested in the financial success of a company; creditors are also concerned with viability. A better firm optimization goal may be the maximization of a firm's total market value, which equals the market value of the equity plus the debt. Usually actions that increase the value of equity will also increase the value of the debt, but this is not always the case. Consider a situation where an action by the firm causes the value of the debt to fall, but the stock price rises. This action has caused some of the value to be transferred from the creditors to the stockholders, but this is considered unethical. All future debt obligations will take into account this prior action through higher interest rates or more restrictive bond covenants. If the valuation or rating of the target and bidder firms' bonds decreases after a pharmaceutical merger, returns have been shifted away from the bondholders to other stakeholders. A merger that exhibits efficiency gains will partially be demonstrated in bond ratings that increase or are unchanged.

Customers are another important class of stakeholder. More and better products increase the choices available to the utility-maximizing consumers. Customers determine part of the success of a merger. If the new integrated firm produces products demanded by the consumers, the new firm's value should increase. For pharmaceutical firms, sales are determined by the interaction of final consumers and their agents. Few final consumers determine product choice. Instead, the agents determine which drug, which manufacturer, and at what price the patient may consume. The manufacturing firms realize the power the agents have over sales and market share. Therefore, in order to maximize stakeholder wealth, the drug manufacturers must consider the opinions and actions of physicians, pharmacies, hospitals, and managed care organizations to the merger.

Employees of the firms constitute the fourth stakeholder. They seek job security and increases in pay. While the stockholders own the firm, the employees are the ones who carry out the day-to-day operations and make most of the decisions. As noted in the industrial organization literature, principal-agent issues can arise when the owning and decision-making processes are separated. After a merger, it is the employees who undertake the integration process. Because employees are instrumental to the success of a merger, their actions and opinions must be considered by the pharmaceutical firms in order to realize wealth gains.

A fifth group of stakeholders includes the government and society in general. They are interested in an increase in the overall wealth of the country. A society may become wealthier if it has more choices, and a firm promotes wealth if it provides more choices. However, social welfare does not always rise when choices increase because the opportunity cost of the increase may offset the benefit of more choice. Hence, an increase in choice can but does not always increase social welfare. Society also wants the general standard of living to increase. For the pharmaceutical industry, this can come from better drugs that produce a healthier life along with strong financial returns, which provide income to owners, creditors, and employees. Society may be interested in drugs that cure ailments that affect a few people and for people who cannot afford drugs. Therefore, pharmaceutical mergers that increase concentration may be acceptable if the firms innovate new drugs at affordable prices. Because the government can prohibit the merger, the firms must note society's desires.

The stakeholders are tied together in the valuation process. The consumers along with their agents determine sales. Employees partially determine costs and product choices. The government regulates the industry and can pressure the firms into conducting business as it wishes. These three stakeholders affect the cash flows for the firm, which determine the stock price and bond value, and which link them to the stockholders and bondholders. When two firms are considering a merger and throughout the integration process, the opinions and actions of all stakeholders need to be considered, for they all affect the new firm's wealth. However, the concerns of the stakeholders may not weigh evenly in the firm's decisions. But under an assumption of efficiency, the firm that makes the best decisions should see its value increase in comparison to its competitors.

Integration Considerations

The main reasons for mergers are to achieve efficiencies and to take advantage of monopoly-related conditions. Efficiencies come in two forms: technical savings, either physical or organizational, and reduction in transactions costs. The monopoly incentives can be from raising the barriers to entry and vertical squeezes. The net social effects of a merger include achieving a balance between net economies gained by a merger that could not be obtained by direct growth or long-term contracts and possible anticompetitive effects such as from raising barriers to entry. For the pharmaceutical industry, the benefit of the mergers may come in the form of increased products and decreased costs. These benefits must exceed the costs of increased concentration and possible increased pricing power to net a positive social gain.

However, mergers that increase concentration do have benefits. Competition can be shallow and myopic. Instead of focusing on the long run, firms focus on a series of short-term contests. Firms become pressured by short-term financial success, which can be in conflict with long-run objectives, if all of the effort is put into just surviving and little is left for creativity. One of the reasons espoused by the pharmaceutical firms for merging is that it allows the firms to have a less myopic view of business because of increased research staffs and cash flow from the increase in critical mass.

Some mergers may cause pecuniary economies, which provide money benefits without improving the use of real resources. Merged firms may be able to get lower input prices, which is a possible result of the vertical mergers between pharmaceutical manufacturers and pharmacy benefit managers. The newly combined firm may also receive promotional advantages from the increased marketing power since sales networks are transferable. Merged pharmaceutical firms can combine detail forces and market more goods. However, if no new innovations or improvements are discovered or no decrease in price results, the benefits of an increase in concentration are limited for consumers.

Mergers may increase diversification through product extension (new product), geographic extension (same product, new location), or pure diversification (no relation to firm's products). Product extension and geographic extension diversification may be beneficial to all stakeholders. However, the benefit of pure diversification is questionable for stockholders and creditors. It may be more efficient for investors to seek diversification by investing in different firms instead of one firm investing in different areas. Employees are unable to diversify on their own and clearly benefit from diversification. Because pharmaceutical firms typically specialize in several therapeutic areas, a merger can expose a firm to new markets. The pharmaceutical firms also tend to have strong ties in various countries. After a merger, existing products have potential new geographic markets.

The total effects of a merger require time before they are realized. Firms may not immediately gain from the external growth, but instead only after a successful integration. The success can then be passed on to employees, consumers, and other stakeholders in the form of more and better products, lower prices, and higher wages.

Returns to Stockholders and the Firm

The success of the merger can be judged in two ways. The first is the return to investors. The stockholders' return is calculated as $R = (P_t + D_t) \div P_{t-1}$ where

R = return on a stock
P_t = selling price in time period t
P_{t-1} = purchase price in period $t - 1$
D_t = dividend per share in period t.

If the owner did not buy or sell the stock during the time period, the return is found by estimating the arithmetic mean of the high and low price per share in period t and in period $t-1$. The return over several time periods is found by taking the geometric mean of each period's return. The geometric mean is used because it is a more conservative average and better suited for outliers. The return is dependent on the dividends received and the change in price or potential change in price. If investors believe that the merger is advantageous, they will bid up the price until the discounted expected future return equals the current selling price. After the merger occurs, the price will continue to increase if the returns from the merger are greater than expected. If the price falls, this indicates that the views of the benefits of the merger were overzealous. If the price remains the same, the merger accomplished what investors expected.

A second measure to judge a merger is to examine the return on assets, which indicates firm performance. The change in the stock price before the merger indicates what the market believes will occur. The change in return on assets shows what did occur. If the merger is successful, then the stock price will continue upward at a quicker pace. However, if the merger is not successful, the stock price will fall. The performance measure will not improve after an unsuccessful merger. It indicates the success of the integration process. The return on assets demonstrates how well the firm utilizes its assets to produce sales and earnings.

Stock price returns and company performance are the two key ways to analyze a merger. Both measures indirectly include the opinions of other stakeholders though their effects on sales, cash flows, bond covenants, and regulations. Direct effects on consumers and bondholders is an additional way to judge the success of a merger. Returns to consumers can be examined by analyzing the number and type of new drugs innovated. After a successful merger, the new firm should produce more and better drugs for its consumers. Bond ratings demonstrate the effect of the merger on bondholders. If the ratings improve, bondholders receive a gain because their bonds' coupon rate is higher than it should be because it was based on the lower-liquidity premerger firm.

Methods and Procedures:

The focus of the empirical section of this study is on returns to economic agents. Economic and finance theories examine mergers over a wider time frame than do the existing empirical studies. This study combines the event study process of the existing studies with the longer view of the current theory. This is accomplished by examining the pharmaceutical industry alone, which will remove any industry effects, and by using a time frame of two years, which captures more of the full effects of the integration process and decreases the effects from external factors. The data are monthly.

Synergy was the stated objective of the mergers by the firms: The horizontal mergers allowed for increased research and development and marketing, while the vertical mergers allowed for some influence over prescription insurance choices and pricing. The firms presented the mergers as a strategic move to counter the pressures of R&D risk and external cost controls. By increasing firm size, the manufacturers wanted to increase control over their environment and increase their ability to service their stakeholders. This research will attempt to infer whether the stated objectives for the mergers were realized. The absolute value and returns to various stakeholders infers the existence and magnitude of synergies or of agency problems. Efficiency gains are shown by nonnegative results whereas nonpositive outcomes indicate agency problems or imperfect information regarding potential synergies.

The first empirical analysis will focus on returns to the various stockholders of firms that have merged. Using an event study, the stock price returns will be investigated with two parametric models and one nonparametric model. The three models allow for an assessment of the returns to bidder and target stockholders. The mean adjusted return model is the historical method, but the market model includes a measure for risk. The nonparametric sign test will be used to test the consistency of the parametric results. In a second empirical analysis, regression will be employed to investigate the annual returns of the mergers on firm performance. In a third empirical analysis, the number and type of new drugs innovated for consumers will be investigated.

Returns to Stockholders

An event study measures the impact of a specific event on the value of a firm. As long as the market is rational, the effect of an event will be reflected in the price of the firm's stock. The first step in constructing an event study is to determine the time frame for the study. Prior studies that used a narrow window around the event (the merger announcement) measure the expectations about the merger but not necessarily the actual merger result. By widening the time frame, the impact of the integration process is included. While the possibility of the inclusion of effects other than the merger increases as the time frame widens, a wider time frame is needed in order to assess whether the initial expectations about the merger are actually fulfilled. Passive management, better known as the buy-and-hold method, advocates long-term possession of stocks instead of repetitive and frequent buying and selling of stocks. This is because ownership of a stock is optimally based on the fundamentals of the firm and not temporary movements in the stock price. Also, taxes and commissions diminish the gains from frequent buying and selling. Therefore, while the investor is interested in the immediate stock price effect of a merger announcement, the long-term effect of the merger is more relevant for portfolio theory. This study will use a time frame of two years around the merger announcement.

The event study will use a concept called the abnormal return in its measurement. The abnormal return is the difference between the actual return and the normal return. The normal return is the predicted return that would be expected if no event occurred. For firm i and event date t the abnormal return is $AR_{it} = R_{it} - E(R_{it}|X_t)$, which is the actual return for the firm minus the predicted return. The abnormal return, also called the residual, is an estimate of the change in the firm value attributed to the event, in this case the merger. The abnormal returns are then summed to produce a cumulative abnormal return, which is used in the test statistic. There are two main methods for determining the normal return: the mean adjusted return model and the market model. Both methods use a clean period. During the clean period, no information about the merger is released. In this study, the clean period will be the third and fourth years prior to the event.

In the mean adjusted return method, the average return for the firm is calculated during the clean period,

$$\hat{R} = \sum_{t=1}^{n} R_{it}/n,$$

where n is the number of observations. The average return is used as the predicted normal return in the calculation of the abnormal return.

The market model is $\hat{R}_{it} = \alpha_i + \beta_i R_{mt} + e_{it}$. R_{mt} is the return on the market index for time period t, β_i is a measure of risk of the firm and α_i measures the variability not explained by the market. The market model is estimated by running a regression on the clean period. The regression yields estimates of α_i and β_i. The normal return for the event period is found by substituting the estimates into the equation with the actual return for the market for the time period. Weston, Chung, and Siu (1998) and MacKinlay (1997) consider this to be the most widely used method because it includes a measure for risk even though the results are comparable to the results from the mean adjusted method.

After obtaining estimates of the normal return, the abnormal return is calculated. Residual analysis tests whether the actual return to the stock is greater or less than what would be normally expected from the basic risk versus return relationship. The abnormal returns for the firm are aggregated across time to arrive at a cumulative abnormal return. The cumulative abnormal return (CAR) is assumed to be normally distributed.

$$CAR_i(t_1,t_2) = \sum_{t=t_1}^{t_2} AR_{it}$$

and

$$CAR_i(t_1,t_2) \sim N(0, \sigma_i^2(t_1,t_2))$$

(MacKinlay, 1997). The CAR is then divided by its standard deviation to arrive at a t-statistic, which can then be tested for significance across the event time period.

MacKinlay (1997) recommends the use of both parametric and nonparametric tests. The nonparametric test can check the robustness of parametric results and may provide more reliable inferences depending on the data set. The nonparametric test used to analyze abnormal returns is the sign test. A benefit of this test is that it does not assume a specific distribution of the abnormal returns. The sign test is based on the sign of the abnormal return. It requires that the residuals are independent and have an expected proportion of positive values equal to 0.5. The null hypothesis is that it is equally probable that the ARs will be positive or negative. If the null hypothesis is rejected, a positive or negative trend exists in the abnormal returns. The total number of cases (N) and the total number of positive cases (N^+) are needed to calculate the test statistic (Conover, 1980; MacKinlay, 1997). The test statistic ($\hat{\theta}$) is,

$$\hat{\theta} = \left[\frac{N^+}{N} - 0.5\right] \sqrt{N} / 0.5 \sim N(0,1).$$

These procedures will allow for inferences about the returns to stockholders, both bidder and target. The event studies examine whether the stockholders gain significant positive or negative returns from the merger event.

Operating Performance of the Firms

Annual returns to the firms for the period between 1984 and 1997 will be examined using regression analysis. Returns to the firm are measured by the return on assets. The return on assets is computed by dividing net income by total assets. It is also the product of the total asset turnover and the net profit margin. The return on assets examines how efficiently a firm is using its assets to generate sales and earnings. The results of the regression analysis will allow inferences about the effects of the mergers on operating performance.

Factors that may impact the returns to economic agents are the return on the S&P 500 index, market return, long-term debt, cash flows, the type of merger, the year of observations, sales, firm classification, and a merger time variable. The S&P 500 index will be used to account for overall movements in the returns to stocks due to market conditions. The type of merger is defined as a horizontal or vertical merger. Year of the observations will be used to capture a potential trend. Annual sales will account for the size of each firm and will be used to assess if the motivations for the mergers were realized. Firms are classified as either a bidder or target firm. The merger time variable refers to whether the yearly observations are for a firm before or after it merges. Firms for this analysis include the following: American Home Products, American Cyanamid, Bristol-Myers, Squibb, Merck, Medco, and Eli Lilly. Consistent with empirical work in finance, first difference and annual growth data transformations will be employed to correct for potential time-series autocorrelation problems.

Returns to Consumers and Bondholders

Mergers also affect consumers and bondholders. The number and type of new drugs innovated implies a return to consumers. A drug is classified by the Food and Drug Administration based on its scientific gain. The postmerger innovations will be examined to determine if they provide a significant gain in therapeutic value versus the premerger new drugs and thus a gain to consumers.

The bond ratings pre- and postmerger will be examined to determine the return to bondholders. If a firm has a decrease in liquidity, the bond ratings will fall, causing a loss to bondholders. The movement of the ratings will allow for inference as to the gains or losses incurred by the bondholders.

Data and Sources

The data requirements for the events studies include stock prices and dividends for each firm and the S&P 500 index. Standard and Poor's *Daily Stock Price Record* records this daily information about each firm traded in the United States and the index.

Firm-level data needed for returns to the firm include assets, net income, sales, cash flows, long-term debt, and equity. The data will be obtained from the annual 10-K reports submitted to the Securities and Exchange Commission (SEC) available on the Internet from the EDGAR database (www.sec.gov/edgar) and from Lexus-Nexus Company Financial Reports (www.lexus-nexus). All publicly traded firms must submit a 10-K report to the SEC each year that summarizes the financial activity of the firm. Specific accounting guidelines must be followed in the 10-K report. The consumers price index (CPI) to deflate dollar denominated variables like long-term debt, cash flow, and sales will be obtained from the Bureau of Labor Statistics.

In evaluating returns to consumers, information about the number of drugs and types of drugs is available from the Food and Drug Administration. The FDA also ranks each drug for its therapeutic value. Bond-quality ratings are available from Moody's and Standard and Poor's.

References:

Acs, Zolton J., and David B. Audretsch. "Innovation in Large and Small Firms: An Empirical Analysis." *American Economic Review*, Vol. 78 (September 1988): 678–90.

Agrawal, Anup, Jeffrey F. Jaffe, and Gershon N. Mandelker. "The Post-Merger Performance of Acquiring Firms: A Re-examination of an Anomaly." *Journal of Finance*, Vol. 47 (September 1992): 1605–21.

Arrow, K. J., "Vertical Integration and Communication." *Bell Journal of Economics*, Vol. 6 (Spring 1975): 173–83.

Berkovitch, Elazar, and M. P. Narayanan. "Motives for Takeovers: An Empirical Investigation." *Journal of Financial and Quantitative Analysis*, Vol. 28 (September 1993): 347–62.

Boston Consulting Group. *The Changing Environment for U.S. Pharmaceuticals.* Boston: Boston Consulting Group, 1993.

Bradley, M., A. Desai, and E. H. Kim. "Synergistic Gains from Corporate Acquisitions and Their Division between the Stockholders of Target and Acquiring Firms." *Journal of Financial Economics*, Vol. 21 (1988): 3–40.

Brigham, Eugene F., and Louis C. Gapenski. *Intermediate Financial Management*, 4th ed. New York: Dryden Press, 1993.

Capen, E. C., R. V. Clapp, and W. M. Campbell. "Competitive Bidding in High-Risk Situations." *Journal of Petroleum Technology*, Vol. 23 (June 1971): 641–53.

Caves, Richard E., Michael D. Whinston, and Mark A. Hurowitz. "Patent Expiration, Entry, and Competition in the U.S. Pharmaceutical Industry." *Brookings Papers: Microeconomics 1991* (1991): 1–66.

Comanor, William S. "The Political Economy of the Pharmaceutical Industry." *Journal of Economic Literature*, Vol. 24 (September 1986): 1178–1217.

Dess, G. G., R. D. Ireland, and M. A. Hitt. "Industry Effects and Strategic Management Research." *Journal of Management*, Vol. 16, No. 1 (1990): 7–27.

DiMasi, Joseph A., Ronald W. Hansen, Henry G. Grabowski, and Louis Lasagna. "Cost of Innovation in the Pharmaceutical Industry." *Journal of Health Economics*, Vol. 10 (1991): 107–42.

Fama, E. F., and M. C. Jensen. "Organizational Forms and Investment Decisions." *Journal of Financial Economics*, Vol. 14 (April 1985): 101–19.

Grabowski, Henry G. "An Analysis of US International Competitiveness in Pharmaceuticals." *Managerial and Decision Economics*, Special Issue (Spring 1989): 27–33.

Grabowski, Henry G., and John M. Vernon. "Brand Loyalty, Entry, and Price Competition in Pharmaceuticals after the 1984 Drug Act." *Journal of Law and Economics*, Vol. 35 (October 1992): 331–50.

———. "A New Look at the Returns and Risks to Pharmaceutical R&D." *Management Science*, Vol. 36 (July 1990): 804–21.

———. "Returns to R&D on New Drug Innovations in the 1980s." *Journal of Health Economics*, Vol. 13 (1993): 383–406.

Greer, Douglas F. *Industrial Organization and Public Policy*. New York: Macmillian, 1992.

Health Care Financing Administration, Office of Managed Care. *Managed Care Enrollment Report, 1995*. Washington, D.C.: Government Printing Office, 1995.

Healy, Paul M., Krishna G. Palepu, and Richard S. Ruback. "Does Corporate Performance Improve after Mergers?" *Journal of Financial Economics*, Vol. 31 (1992): 135–75.

Houlihan, Lokey, Howard & Zukin. *Mergerstat Review*. Los Angeles: 1997.

Jarrell, G. A., J. A. Brickley, and J. M. Netter. "The Market for Corporate Control: The Empirical Evidence since 1980." *Journal of Economic Perspectives*, Vol. 2 (Winter 1988): 49–68.

Jensen, M. C. "Agency Costs of Free Cash Flow, Corporate Finance, and Takeovers." *American Economic Review*, Vol. 76 (May 1986): 323–29.

Jensen, M. C., and R. S. Ruback. "The Market for Corporate Control: The Scientific Evidence." *Journal of Financial Economics*, Vol. 11 (1983): 5–50.

Keown, Arthur, J. William Petty, David F. Scott, Jr., and John D. Martin. *Foundations of Finance*, 2nd ed. Upper Saddle River, NJ: Prentice-Hall, 1998.

Klein, B., R. Crawford, and A. Alchian. "Vertical Integration, Appropriate Rents, and the Competitive Contracting Process." *Journal of Law and Economics*, Vol. 21 (October 1978): 297–326.

Lu, Z. John, and William S. Comanor. "Strategic Pricing and New Pharmaceuticals." Mimeo, University of California at Santa Barbara, 1994.

Lubatkin, M., and R. E. Shrieves. "Towards Reconciliation of Market Performance Measures to Strategic Management Research." *Academy of Management Review*, Vol. 11, No. 2 (1986): 497–512.

MacKinlay, A. Craig. "Event Studies in Economics and Finance." *Journal of Economic Literature*, Vol. 35 (March 1997): 13–39.

Mitchell, Mark L., and J. Harold Mulherin. "The Impact of Industry Shocks on Takeover and Restructuring Activity." *Journal of Financial Economics*, Vol. 41 (June 1996): 193–229.

Moyer, R. Charles, James R. McGuigan, and William J. Kretlow. *Contemporary Financial Management*, 5th ed. New York: West Publishing Company, 1990.

Myers, Stewart C., and Nicholas J. Majluf. "Corporate Financing and Investment Decisions When Firms Have Information That Investors Do Not Have." *Journal of Financial Economics*, Vol. 13 (June 1984): 187–221.

Office of Technology Assessment. *Pharmaceutical R and D: Costs, Risks, and Rewards*. OTA-H-522. Washington, D.C.: Government Printing Office, February 1993.

PhRMA (Pharmaceutical Research and Manufacturers of America). *Annual Survey*. 1997a. Available from http://www.pharma.org/index.html; Internet; accessed December 16, 1997.

———. *Industry Profile*. 1997b. Available from http://www.pharma.org/facts/industry/index.html; Internet; accessed December 16, 1997.

Roll, Richard. "The Hubris Hypothesis of Corporate Takeovers." *Journal of Business*, Vol. 59 (April 1986): 197–216.

Scherer, F. M. "Pricing, Profits, and Technological Progress in the Pharmaceutical Industry." *Journal of Economic Perspectives*, Vol. 7 (Summer 1993): 97–115.

Schleifer, Andrei, and Robert Vishny. "Large Shareholders and Corporate Control." *Journal of Political Economy*, Vol. 94 (June 1988): 461–88.

Schweitzer, Stuart O. *Pharmaceutical Economics and Policy*. New York: Oxford University Press, 1997.

Subramanian, R., E. Mahmoud, M. Thibodeaux, and B. Ebrahimi. "Evaluating Merger Performance on a Longitudinal Basis: An Empirical Investigation." *Journal of Business Strategies*, Vol. 9, No. 2 (Fall 1992): 87–100.

Sudarsanam, P. S. *The Essence of Mergers and Acquisitions*. London: Prentice-Hall, 1995.

Travlos, Nicholaos G. "Corporate Takeover Bids, Methods of Payment, and Bidding Firms' Stock Returns." *Journal of Finance*, Vol. 42 (September 1987): 943–63.

Wansley, James W., William R. Lane, and Ho C. Yang. "Abnormal Returns to Acquired Firms by Type of Acquisition and Method of Payment." *Financial Management*, Vol. 12 (Autumn 1983): 16–22.

Weston, J. Fred, Kwang Chung, and Juan Siu. *Takeovers, Restructuring, and Corporate Governance*, 2nd ed. Englewood Cliffs, NJ: Prentice-Hall, 1998.

Williamson, O. E. Markets and Hierarchies: Analysis and Antitrust Implications. New York: Free Press, 1975.

Appendix D

Guidelines for Critiquing Papers

Providing useful, constructive criticism on the writing done by others is a demanding task in most cases. Beyond providing the author with useful observations and suggestions, it also helps the person doing the critique to be more analytical toward his or her own writing. A good critique deals with both the general flow and organization of the paper and the details in the paper. A critique's purpose is to be critical (it attempts to find any of the flaws and weaknesses in the paper), but it should be done constructively (it helps the author produce a better paper). You should be tactful in presenting your criticisms, but do not skirt the problems and issues. Induce the author to *think* about and question what he or she has written. Challenge the author when it is appropriate.

A list of questions and specific directions to help you provide more effective critiques is given below.

1. How clear is the overall thought and meaning of the piece? Does it meet the objective? How effectively and efficiently does it do that? Is it written appropriately for the intended audience? Provide the author with your analysis, perspective, and suggestions.

2. Is the *flow* of thought clear? Does it "guide" the reader's thinking and understanding to the point (objective) that the writer meant to achieve? Is the progression of thought such that it is easy to follow? Are the individual points or thoughts connected and in the most effective order? Provide suggestions to the writer on how to improve it.

3. How well is the written paper organized? Does it use headings, subheadings, and other manuscript divisions well? If not, can you suggest ways to improve it?

4. Are there ambiguous, normative, or confusing words used that convey the wrong impression or that may not make the writer's meaning as clear as possible? Are all words and phrases technically correct? Offer suggestions as well as criticisms.

5. Cover both the general and specific aspects of the paper, but concentrate on the general considerations (e.g., overall organization and flow of thought) first, then deal with the details (e.g., specific wording, grammar, etc.).

6.   Help the writer with corrections and suggestions on straight editorial matters (spelling, punctuation, sentence structure, etc.). However, it is not your function to "clean up" someone else's poor writing. If someone asks you to review a paper, they owe you the courtesy of making sure that the mundane matters have received sufficient attention to make good use of your valuable time. If they have not done so, they do not deserve a helpful, or even courteous, response from you.

The above guidelines and suggestions apply to papers in general. For additional perspective specific to reviewing and critiquing for economics journals, see Hamermesh (1992, pp. 176–179).

Appendix E

Seeking Research Funding

Advice on seeking funding for research seems a foolhardy undertaking; no advice offered can be reliable all the time. However, some perspective on matters of public and private funding of economic research may be instructive and is offered on the premise that your perspective may be expanded in the process.

In the discussion below, thoughts on funding of research are grouped under two headings: dedicated funding and competitive funding. They are treated separately because they are fundamentally different in both structure and intent. Obtaining competitive research funding poses a distinct set of problems, particularly for beginning professional economists.

DEDICATED FUNDING FOR ECONOMIC RESEARCH

Dedicated, sometimes called "hard," funding for research may take several forms. The most obvious source of dedicated support for economic research is employment with organizations that hire economists to conduct research. Many federal and state government departments and agencies employ economists in research positions. The U.S. Departments of Labor, Treasury, Education, and Agriculture, and their counterparts in state governments, are examples. Other organizations such as the Federal Reserve Banks, the National Bureau of Economic Research, utility companies, stock and commodity trading companies, foundations, and consulting firms are also examples of organizations that employ economists to do research. Often the employment provided in these organizations is not for research alone but includes a mixture of research, policy and staff analysis, and management support. Additionally, the research conducted within these types of organizations is predominantly applied (subject-matter or problem-solving) research. The small proportion of disciplinary research from these sources seems to be oriented to analytical techniques more than to theoretical developments, at least in recent years.

The primary source of dedicated funding for disciplinary research in economics may be the employment of economists by universities. This funding is implicit; there is an implied agreement within university structures that we will do research "on our own" (i.e., without funding other than base salary support) that will result in journal publications. Technically, we are paid to teach; but it is understood that we also will do research (and administrative and service functions). The maintained contention is that this type of funding has been the most important *single* source of funding for disciplinary research in the United States during the twentieth century. This means of funding sometimes appears to be under growing pressure and may contribute less research support in the future.

Another, somewhat unique, source of funding for economic research in the United States is the dedicated system of federal funding of land-grant universities by the U.S. Department of Agriculture to support research and extension activities that benefit rural America. Applied economic research, largely problem-solving research, is supported by Hatch Act funds, which came into existence in 1865. While this funding may be largely responsible for the relative abundance of food and fiber in the United States, it remains an anomaly in the system of research support in applied economics. Most economic research is supported as noted above or through competitive research grants. Further, funding from this source has been decreasing in both relative and absolute terms and in both nominal and real terms.

COMPETITIVE FUNDING FOR ECONOMIC RESEARCH

Government departments and agencies, international organizations, commodity and trade organizations and groups, and such institutions fund an array of competitive research grant programs. Obtaining funding from these sources is particularly important for economists in universities and consulting organizations. Universities and consulting organizations invest substantial resources to obtain information about and compete for competitive grants. Many economists find the process to be fraught with problems, but we often have little choice if we intend to conduct research.

Young professionals often become frustrated with the "grantsmanship" game. Some professionals stop trying after a series of unsuccessful efforts; others "burn out" after years of playing the game. The thoughts offered below are directed to young professional economists, and potentially some graduate students, who are confronted with the need to obtain external funding to support their research. For many in that position, establishing a start on external funding is critical to their professional success.

Reviewers of competitive grant proposals often review dozens, and sometimes hundreds, of proposals from a single announced grant program. Reviews often have three components or stages—solicited reviews by experts in the subject-matter field of the proposal, a review panel assembled by the agency, and a "program officer" employed by the agency whose job entails managing the review process. In any case, you need something that distinguishes your proposal, some aspect that captures reviewers' imagination or curiosity and causes them to remember your proposal. Call this a "hook" or a "gimmick" if you wish. Remember, however, that an interesting approach or presentation will not substitute for absence of substance in the proposal. Advocating creative presentation of a good proposal does not translate to gimmickry to compensate for a weak proposal. Reviewers are, in general, too astute not to recognize lack of substance. However, a reviewer reading dozens of proposals can easily overlook a good proposal if something does not capture his or her attention.

Your most effective approach is to develop a well-planned, well-written proposal that provides reviewers with a clear understanding of the proposed research and adheres to the instructions on proposal format and the mission of the agency or organization. Proposals that are longer than necessary, wordy, rambling, and the like often receive poor reviews. Reviewers faced with reading many proposals quickly develop low tolerance for long, unclear proposals and an appreciation for clear, space-efficient ones. Aspects of effective proposals were a major consideration of chapters 5 through 9. It is the author's belief that if a reviewer does not have a clear understanding of the research problem and objectives as they are stated in the proposal, and does not accept the objectives, the remainder of the proposal will not receive the benefit of the doubt in the reviewer's mind. The reviewer may not even read any further or may continue with diminished attention.

Given good research ideas and effective presentation of them, recognize and accept the probabilistic or risk elements in the process. It will often help your case to identify potential pitfalls in the research and discuss alternative ways to deal with them. We do not have reliable data on probability of success with research proposals in economics, but the success rate for *professional* proposals at land-grant universities in the United States across all disciplines is 45% (Neuen-schwander, 1992, editor's note). The success rate for the beginning professional will be lower. Perseverance is an important attribute for success. Consequently, seeking competitive funding for research is not for persons who are easily discouraged or persons without a degree of philosophical tolerance for rejection. In some respects, it is not unlike being in graduate school permanently; you must repeatedly demonstrate your understanding, capability, and insight.

The following summary list of suggestions for developing grant-funded proposals draws from several sources as well as personal experience.

1.   Know the funding agency and its mission, and do not hesitate to ask questions of the program officer who is managing the competitive grants program.

2.   Follow the instructions/guidelines in the Request for Proposals (RFP) carefully, and pay attention to the details (format, communication, and technical writing details). Edit thoroughly and prepare to write and rewrite.

3.   Establish the focus and relevance of your proposed research quickly (page 1) and establish how the research relates to the larger scientific issues. This process may be fostered by submitting early since proposals are often reviewed in the order received; yours may receive more attention if it avoids the "rush."

4. Establish your credentials for conducting the research, which will be influenced by the effectiveness of the literature review.
5. Make the expected products/impacts of the research clear.
6. The budget should be realistic and appropriate for the work proposed.

For additional reading on securing grant funding, suggestions include Gangler and Freckman (1990), Neuenschwander (1992), and *Science* (1994).

As you gain experience in writing proposals and as you experience some success with funding and research projects, the process will become somewhat easier and your success rate will increase. As that occurs, you are building a record or reputation for doing accurate, reliable, useful research. If you can build that reputation, funding becomes relatively easier to obtain. Given that much of our research funding must come from sources that are concerned with subject-matter and problem-solving matters, it may not be sufficient to have a good research reputation solely within the economics discipline. Reviewers of competitive grants proposals often include economists and noneconomists from academia, government, and industry. Consequently, your reputation for conducting reliable research that is relevant, done on time, and so on must span both disciplinary and nondisciplinary areas. Recognize that there may be instances in which reviewers will not have the credentials to judge the disciplinary content of your research proposals, particularly when problem-solving research is being reviewed. You may lament that fact, but your funding success will increase if you accept it.

In the process of getting started as a young professional, it is often advisable to work in conjunction with other researchers with more experience who already have established some reputation for success. This frequently occurs when graduate students continue some cooperative research and/or writing with their major professors or other faculty after receiving their degrees. Working with senior researchers is not limited to one's committee. One or more mentors from the ranks of your fellow professionals early in your career can be a distinct advantage. You may obtain funding more readily and thus begin to establish your own "track record." You probably will not want to "stand in the shadow of" your senior colleague(s) indefinitely, but this symbiotic cooperation can be useful in starting a career. These relationships can be productive for senior researchers as well. Sometimes this kind of cooperative arrangement yields an effective combination of current technical expertise, experience, and perspective in the analytical research process.

Consider the observation that money may be more readily available for multidisciplinary and interdisciplinary research in contrast to single-discipline, subject-matter, and problem-solving research. It seems that funding organizations and groups believe or recognize that many "real-world" issues and problems cannot be adequately addressed within a single discipline. Should you choose, or be in a position, to capitalize on opportunities in multidisciplinary research efforts, be forewarned that successful efforts of this nature may not be as simple as you assume. Doing team research requires commitment from each of the disciplinary members as a minimum requirement. Also, it likely requires a substantial commitment of time for communication between or among the disciplines. The communication problems of understanding each other's disciplinary language is, in the author's opinion, the greatest problem in conducting research across disciplines. The benefits, both monetary and personal satisfaction, can be rewarding, however. The experience can also be painful if not managed well.

A note of caution: When confronted with "opportunities" to obtain funding in exchange for producing "research results" that support a predetermined position, resist the temptation, particularly if you want to be taken seriously as a researcher. This can be a problem for economists working in multidisciplinary teams when the physical scientists or engineers want the technology or proposed innovation to "look economically favorable." The marketability of your analytical talents within the research community will be limited unless you are perceived as an objective individual. A reputation as a "hired gun" can decimate an otherwise promising research reputation. Consulting work carries a poor image among some researchers, perhaps because it sometimes assumes the position of special pleading (a logical fallacy discussed in chapter 3), irrespective of the content of the analysis. Young professionals are well advised to be particularly careful to have a clear understanding with, for example, private and government funding groups, regarding research results "falling where they may" and publication of the results. The author's personal experience on these matters has been that problems are avoided if the agreements regarding objectivity, publishing results, and such matters are dealt with explicitly in the research proposal stage.

References

Alley, Michael. *The Craft of Scientific Writing*. Englewood Cliffs, NJ: Prentice-Hall, Inc., 1987.

Andrew, Chris O., and Peter E. Hildebrand. *Planning and Conducting Applied Agricultural Research*. Boulder, CO: Westview Press, 1982.

Ary, Donald, Lucy Cheser Jacobs, and Asghar Razavieh. *Introduction to Research Education*. 2nd ed. New York: Holt, Rhinehart and Winston, 1979.

Austin, James H. *Chase, Chance, and Creativity: The Lucky Art of Novelty*. New York: Columbia University Press, 1978.

Ayala, Francisco, Robert McCormick Adams, Mary-Dell Chilton, Gerald Holton, David Hull, Kumar Patel, Frank Press, Michael Ruse, and Phillip Sharp. *On Being a Scientist*. Committee on the Conduct of Science, National Academy of Sciences. Wash., DC: National Academy Press, 1989.

Babbie, Earl. "New Student Guidelines for Plagiarism via the World Wide Web." http://www.chapman.edu/wilkinson/socsci/sociology/Faculty/Babbie. 1998.

Back, W. B. Unpublished notes for a course in research methodology at Oklahoma State University, no date (probably in the 1960s).

Bessler, David A. "Quantitative Techniques: A Discussion." *Agriculture and Rural Areas Approaching the Twenty-first Century*, R. J. Hildreth, Kathryn L. Lipton, Kenneth C. Clayton, and Carl C. O'Connor, editors. Ames, IA: Iowa State University Press, 1988, pp. 199–204.

Blaug, Mark. *The Methodology of Economics, or How Economists Explain*. New York: Cambridge University Press, 1980.

Bloom, Benjamin S., Max D. Engelhart, Edward J. Furst, Walker H. Hill, and David R. Krathwohl. *Taxonomy of Educational Objectives: The Classification of Educational Goals*. Handbook I: Cognitive Domain. New York: David McKay Co., Inc., 1956.

Boland, Lawrence A. *The Foundations of Economic Method*. London: George Allen & Unwin, 1982.

Bonnen, James T. "A Century of Science in Agriculture: Lessons for Science Policy." *American Journal of Agricultural Economics*. 68 (1986): 1065–1080.

Breimyer, Harold F. "Scientific Principle and Practice in Agricultural Economics: An Historical Review." *American Journal of Agricultural Economics*. 73 (1991): 243–254.

Brewster, John M. *A Philosopher among Economists*. J. Patrick Madden and David E. Brewster, editors. Philadelphia: J. T. Murphy, 1970.

Brorsen, Wade. "Observations on the Journal Publication Process." *North Central Journal of Agricultural Economics*. 9 (1987): 315–321.

Buffington, Perry W. "In Our Right (and Left) Minds." *Sky*. (1986): 33–34.

Caldwell, Bruce J. *Beyond Positivism: Economic Methodology in the Twentieth Century*. London: George Allen & Unwin, 1982.

Castle, Emery N. "Economic Theory in Agricultural Economics Research." *The Journal of Agricultural Economics Research*. 41 (1989): 3–7.

Churchman, C. West. *The Theory of Experimental Inference*. New York: Macmillan, 1948.

236

Colander, David. "The Sounds of Silence: The Profession's Response to the COGEE Report." *American Journal of Agricultural Economics*. 80 (Aug. 1998): 600–607.

Colander, David, and Arjo Klamer. "The Making of an Economist." *Journal of Economic Perspectives*. 1 (1987): 95–111.

Committee on Science, Engineering, and Public Policy. *On Being a Scientist: Responsible Conduct in Research*. 2nd ed. Wash., DC: National Academy Press, 1995.

Daniels, A.M. *Business Information Sources*. Berkeley: University of California Press, 1985.

Edwards, Clark. "Doing Agricultural Economics." *The Journal of Agricultural Economics Research*. 42 (1990): 3–7.

Edwards, Clark. "The Potential for Economics Research." *Agricultural Economics Research*. 30 (1978): 29–35.

Eichner, Alfred S. "Can Economics Become a Science?" *Challenge*. 29 (1986): 4–12.

Einstein, Albert. "Science and Religion." *The World Treasury of Physics, Astronomy, and Mathematics*, Timothy Ferris, editor, Toronto: Little, Brown, and Co., 1991, pp. 828–835.

Feynman, Richard P. *The Pleasure of Finding Things Out: The Best Short Works of Richard P. Feynman*. Cambridge, MA: Perseus Publishing, 1999.

Fowler, H. Ramsey. *The Little, Brown Handbook*. 2nd ed. Boston: Little, Brown and Co., 1983.

Freedman, Craig. "Why Economists Can't Read." *Methodus*. 5 (June 1993): 6–23.

French, Charles E. "Agricultural Economics, 1971–1986, or Some Nasty Illustrations of What You Are Getting Yourself Into." Presentation to the International Graduate Seminar, Purdue University, Aug. 12, 1971.

Friedman, Milton. "The Methodology of Positive Economics." *Essays in Positive Economics*. Chicago: University of Chicago Press, 1953, pp. 3–43.

Gangler, Randy, and Diana W. Freckman. "A Program Officer's Guide to Effective Grantsmanship." *American Entomologist*. (Fall 1990): 206–212.

Ghebremedhin, Tesfa, and Luther Tweeten. *Research Methods and Communication in the Social Sciences*. Westport, CT: Praeger, 1994.

Gilbert, Susan. "Profiting from Creativity." *Science Digest*. 94 (1986): 37–39, 78, 79.

Goode, William J., and Paul K. Hatt. *Methods in Social Research*. New York: McGraw Hill Book Co., 1952.

Griliches, Zvi. "Hedonic Price Indexes for Automobiles: An Econometric Analysis of Change." *The Price Statistics of the Federal Government*. New York: National Bureau of Economic Research, General Series, No. 73, 1961, pp. 137–196.

Hamermesh, Daniel S. "The Young Economist's Guide to Professional Etiquette." *Journal of Economic Perspectives*. 6 (1992): 169–179.

Hansen, W. Lee. "The Education and Training of Economics Doctorates: Major Findings of the Executive Secretary of the American Economic Association's Commission on Graduate Education in Economics." *Journal of Economic Literature*. XXIX (1991): 1054–1087.

Hausman, David M. "Economic Methodology in a Nutshell." *Journal of Economic Perspectives*. 3 (1989): 115–127.

Heady, Earl O. "Implications of Particular Economics in Agricultural Economics Methodology." *Journal of Farm Economics*. 31 (1949): 837–850.

Heisenberg, Werner. "Positivism, Metaphysics, and Religion." *The World Treasury of Physics, Astronomy, and Mathematics*, Timothy Ferris, editor. Toronto: Little, Brown, and Co., 1991, pp. 821–827.

Indiana University. Plagiarism statement from the Indiana University Code of Student Affairs, 1998.

Johnson, Glenn L. "Disciplines, Processes, and Interdependencies Related to the Problem-Solving and Issue-Oriented Work of Agricultural Economists." Chapter 9, *Beyond Agriculture and Economics: Management, Investment, Policy, and Methodology,* A. Allan Schmid, editor. East Lansing, MI: Michigan State University Press, 1997.

Johnson, Glenn L. "Economics and Ethics." *The Centennial Review.* XXX (Winter 1986a): 69–108.

Johnson, Glenn L. "Holistic Modeling of Multidisciplinary, Subject Matter, and Problem Domains." Chapter 5, *Systems Economics,* Karl A. Fox and Don G. Miles, editors. Ames, IA: Iowa State University Press, 1987.

Johnson, Glenn L. "Multidisciplinary, Problem-Solving, and Subject-Matter Work." *Agricultural Economics.* 5 (1991): 187–192.

Johnson, Glenn L. "Philosophic Foundations: Problems, Knowledge, and Solutions." *European Review of Agricultural Economics.* 3 (1976): 226.

Johnson, Glenn L. *Research Methodology for Economists: Philosophy and Practice.* New York: Macmillan Publishing Co., 1986b.

Johnson, H. G. *On Economics and Society.* Chicago: University of Chicago Press, 1975, 129–139.

Johnson, S. R. "Quantitative Techniques." Chapter 5, *Agriculture and Rural Areas Approaching the Twenty-first Century,* R. J. Hildreth, Kathryn L. Lipton, Kenneth C. Clayton, and Carl C. O'Connor, editors. Ames, IA: Iowa State University Press, 1988, pp.177–198.

Keynes, John Maynard. *General Theory of Employment, Interest and Money.* New York: Harcourt Brace, 1936.

King, Richard A. "Choices and Consequences." *American Journal of Agricultural Economics.* 61 (1979): 839–859.

Krueger, Anne O., Kenneth J. Arrow, Oliver Jean Blanchard, Alan S. Blinder, Claudia Goldin, Edward E. Leamer, Robert Lucas, John Panzar, Rudolph G. Penner, T. Paul Schultz, Joseph E. Stiglitz, and Lawrence H. Summers. "Report of the Commission on Graduate Education in Economics." *Journal of Economic Literature.* XXIX (1991): 1035–1053.

Lacey, A. R. *A Dictionary of Philosophy.* London: Routledge & Kegan Paul Ltd., 1976.

Ladd, George W. *Imagination in Research: An Economist's View.* Ames, IA: Iowa State University Press, 1987.

Ladd, George W. "Thoughts on Building an Academic Career." *Western Journal of Agricultural Economics.* 16 (1991): 1–10.

Ladd, George W. "Value Judgements and Efficiency in Publicly Supported Research." *Southern Journal of Agricultural Economics.* 15 (1983): 1–7.

Ladd, George W., and Veraphol Suvannunt. "A Model of Consumer Goods Characteristics." *American Journal of Agricultural Economics,* 58 (1976): 504–510.

Lancaster, K. J. "A New Approach to Consumer Theory." *Journal of Political Economy.* 74 (1966): 132–157.

Lansford Publishing Co. "How to Conduct a Survey." San Jose, CA: The Lansford Publishing Co., Inc., 1977.

Lansford Publishing Co. "Questionnaire Construction." San Jose, CA: The Lansford Publishing Co., Inc., 1980.

Larrabee, Harold A. *Reliable Knowledge: Scientific Methods in the Social Studies*. Boston: Houghton Mifflin Co., 1964.

Leedy, Paul D. *Practical Research Planning and Design*. 3rd ed. New York: Macmillan Publishing Co., 1985.

Leontief, Wassily. "Academic Economics." *Science*. 217 (1982): 104–107.

Leontief, Wassily. "Can Economics Be Reconstructed as an Empirical Science?" *American Journal of Agricultural Economics*. 75 (1993, 75th Anniversary Issue): 2–5.

Levins, Richard A. "The Whimsical Science." *Review of Agricultural Economics*. 14 (1992): 139–151.

Levy, J. "Right Brain, Left Brain: Fact or Fiction." *Psychology Today*. 19 (1985): 50–58.

Lewin, Shira B. "Economics and Psychology: Lessons for Our Own Day from the Early Twentieth Century." *Journal of Economic Literature*. XXXIV (Sept. 1996): 1293–1323.

Lewis, Clarence Irving. *The Ground and Nature of the Right*. New York: Columbia University Press, 1955.

Li, Xia, and Nancy B Crane. "Bibliographic Formats for Citing Electronic Information." *Citing Electronic Information*. Meckler, 1996. http://www.uvm.edu/~xli/reference/estyles.html.

Machlup, Fritz. *Methodology of Economics and Other Social Sciences*. New York: Academic Press, Inc., 1978.

Maki, Uskali. "Diagnosing McCloskey." *Journal of Economic Literature*. XXXIII (Sept. 1996): 1300–1318.

Marshall, Alfred. *Principles of Economics*. 8th ed. London: Macmillan & Co., Ltd., 1964.

Maxwell, Nicholas. "What Kind of Inquiry Can Best Help Us Create a Good World?" *Science, Technology, and Human Values*. 17 (Spring 1992): 205–227.

McCloskey, Donald N. "Agon and Ag Ec: Styles of Persuasion in Agricultural Economics." *American Journal of Agricultural Economics*. 72 (1990a): 1124–1130.

McCloskey, Donald N. *If You're So Smart: The Narrative of Economic Expertise*. Chicago: University of Chicago Press, 1990b.

McCloskey, Donald N. "The Rhetoric of Economics." *Journal of Economic Literature*. XXI (June 1983): 481–517.

Merton, Robert K. "The Matthew Effect in Science, II: Cumulative Advantage and the Symbolism of Intellectual Property." *Isis*. 79 (1988): 621.

Mighell, Ronald L., and Elizabeth Lane. "Writing and the Economic Researcher." *Agricultural Economics Research*. 25 (1973): 15–20.

Myrdal, G. *Objectivity in Social Research*. 5th ed. New York: Pantheon Books, 1944.

Neuenschwander, Leon. "Jump Start Your Grant-Winning Potential." *Academic Leader*, Doris Green, editor. Madison, WI: Magna Publications, Inc., May 1992.

Northrop, F. C. S. *The Logic of the Sciences and Humanities*. New York: Meridian Books, 1959.

O'Sullivan, Patrick J. *Economic Methodology and Freedom to Choose*. London: Allen & Unwin, 1987.

Paarlberg, Don. "Methodology for What?" *Journal of Farm Economics*. 45 (1963): 1386–1392.

Page, Melvin E. "A Brief Citation Guide for Internet Sources in History and the Humanities." gophor://h-net.msu.edu:70/00/lists/H-AFRICA/Internet.cit. No date.

Piaget, Jean. *The Construction of Reality in the Child*. New York: Basic Books, Inc., 1954.

Randall, Alan. "Information, Power, and Academic Responsibility." *American Journal of Agricultural Economics*. 54 (1974): 227–234.

Randall, Alan. "What Practicing Agricultural Economists Really Need to Know about Methodology." *American Journal of Agricultural Economics*. 75 (1993, 75th Anniversary Issue): 48–59.

Rosen, Sherwin. "Hedonic Prices and Implicit Markets: Product Differentiation in Pure Competition." *Journal of Political Economy*. 82 (1974): 34–55.

Rosenberg, Elliot M. "When Lessons Hit Home." *Newsweek*, Sept. 19, 1983, p. 11.

Rudner, Richard. "The Scientist *Qua* Scientist Makes Value Judgements." *Philosophy of Science*. 20 (1953): 1–6.

Runes, Dagobert D., editor. *Dictionary of Philosophy*. New York: Philosophical Library, Inc., 1983.

Salter, Leonard A., Jr. "Cross-Sectional and Case Grouping Procedures in Research Analysis." *Journal of Farm Economics*. 24 (1942): 792–805.

Schultz, T. W. "Investment in Human Capital." *American Economic Review* 51 (Mar. 1961): 1–17.

Science. "Grantsmanship: What Makes Proposals Work?" *Science*. 265 (Sept. 1994): 1921–1922.

Segarra, Eduardo. "Simulation versus Mathematical Programming." *Analyzing Dynamic and Stochastic Agricultural Systems,* Michael E. Wetzstein and C. Robert Taylor, editors. Proceedings of a Technical Symposium by the Consortium for Research on Crop Production Systems, Department of Agricultural Economics, Auburn University, Mar. 1991, pp. 35–69.

Simkin, Colin. *Popper's Views on Natural and Social Science*. Leiden, the Netherlands: E. J. Brill, 1993.

Simon, Herbert A. "Rational Decision Making in Business Organizations." *American Economic Review*. 69 (1979): 493–513.

Smith, Dudley T. "The Practice of Research." Presentation to the Texas Agricultural Experiment Station Staff Conference, College Station, TX, Jan. 1983.

Smith, V. L. "Economics of Production from Natural Resources." *American Economic Review*. LVIII (1968): 409–431.

Spencer, Milton H. *Contemporary Economics*. New York: Worth Publishers, 1971.

Stephan, Paula E. "The Economics of Science." *Journal of Economic Literature*. XXXIV (Sept. 1996): 1199–1235.

Stewart, Ian M. T. *Reasoning and Method in Economics: An Introduction to Economic Methodology*. London: McGraw-Hill, 1979.

Stuller, Jay. "The Nature and Process of Creativity." *Sky*. (1982): 38–39.

Taylor, Fred. "Relation between Primary Market Prices and Qualities of Cotton." U.S. Department of Agriculture Bulletin No. 457, Nov. 24, 1916.

Thomson, William. "The Young Person's Guide to Writing Economic Theory." *Journal of Economic Literature*. XXXVII (Mar. 1999): 157–183.

Tomek, William G. "Confirmation and Replication in Empirical Econometrics: A Step toward Improved Scholarship." *American Journal of Agricultural Economics*, 75 (1993, 75th Anniversary Issue): 6–14.

Tweeten, Luther. "Hypothesis Testing in Economic Science." *American Journal of Agricultural Economics*. 65 (1983): 548–552.

Walker, Janice R. "MLA-Style Citations of Electronic Sources." http://www.cas.usf.edu/english/walker/mla.html. No date.

Wasserman, Paul, and Joanne Paskar, editors. *Statistics Sources: A Subject Guide to Data on Industrial, Business, Social, Educational, Financial, and Other Topics for the United States and Internationally.* Detroit: Gale Research Co., 1974.

Waugh, Frederick V. "Quality Factors Influencing Vegetable Prices." *Journal of Farm Economics.* 19 (1928): 185–196.

Webster's New Collegiate Dictionary. Springfield, MA: G. & C. Merriam Co., 1977.

Webster's New World Dictionary of the American Language. College ed. New York: The World Publishing Co., 1968.

Williams, Willard F. Unpublished class notes for a course in research methodology in economics at Texas Tech University, 1984.

Index

Abstractions, 53, 136
Abstracts, 116–117
  dissertation, 117
  research report, 162
Acknowledgments, 170–172
  research report, 162
Advocacy, avoiding, 95
Affirming the consequent fallacy, 46
American Economic Association, 14
Appeal to authority fallacy, 46
Appeal to the people fallacy, 46
Appendices, research report, 165
Argument by analogy fallacy, 47
Assertion, tentative, 132–133
Associations, establishing, 143
Assumptions, 74–75
Attacking the person fallacy, 46
Audience, intended, 94
Authorship, 170–172

Bacon, Francis, 78
Badness, 63–65
Base model analysis, 137
Bayesian procedures, 150
Bibliography, 116–117, 122
Black boxes, 126
Books, reference format, 123
Broder, Joseph, 57
Budget, research, 86, 149
Bulletins, 161
  reference format, 123
Business Information Sources, 151

Causation, 142–144
Census series, 150–151
Charts, 97
Chauvinism, disciplinary, 93
Citations, 97, 170
Clarity, tests of, 48, 63, 67
Classification, 53
Clichés, 95

Coherence
  logical, 70
  test of, 47–48, 63, 67
Commission on Graduate Education in
  Economics, 6, 14
Committee on Science, Engineering, and
  Public Policy, 169
Committee on the Conduct of Science, 6
  plagiarism statement, 121
Communication, effective, 93
Composition fallacy, 47
Computer technology, 3–4
Concept, defined, 129
Conceptual framework
  dissertation proposal, example, 218–222
  guidelines, 136–138
  research proposal, 88, 127–128
  research report, 163–164
  role, 129–130, 138
  secondary benefits, 136, 138
  theory, 131–132
  thesis proposal, example, 197
Conceptualizing, 53, 128, 129
  approach to, 136
Conclusions, 164
  writing, 167–168
Condition, and phenomenon, 142–144
Confirmation, 59–60
  defined, 8–9
Consulting, 235
Control theory, 150
Cooperative Research Information System
  (CRIS), 126
Correlation, 147
Correspondence, tests of, 48, 63, 67
Craft of Scientific Writing, The, 96
Creativity
  applied, 31
  imaginative, 31
  natural, 31
  prescriptive, 31

in research, 30–33
theoretical, 31
Credit, allocation, 170–172
Critiques, of writing, 96, 99, 229–230
Curiosity, 32
Cynicism, 9

Data
accuracy, 151
collection, 17
dissertation proposal, example, 226
economic, 71
forms, 152–153
obtaining, 76–77
primary, 151–152
in problem statements, 108
references, 123
reliability, 47–49, 74
research proposal, example, 187
secondary, 150–151
social, 71
thesis proposal, example, 199
Deduction, 45, 56
scientific approach, 74–77, 78
Descriptors, unnecessary, 107
Dewey, John, 67
Discovery, 59–60
accidental, 17, 35, 84–85
crediting, 169
defined, 8–9
Dissertation
authorship, 173
completeness of, 161
proposal, example, 201–226
Distribution, 147
Division fallacy, 47

Econometrics, 71–72, 144, 155
tools, 147–148
Economic research, functions of theory,
53–54
Economics
agricultural, 35
applied, 3
as art/science, 38–40
disciplinary, 3

institutional, 67
research methodology in, 26–27
Economic theory, functions, 53–54
Economists
agricultural, 155
applied, 12, 92
theoretical, 12, 92
Efficiency, 56
Empiricism, 40, 62, 71–72
history, 142–144
standard, 61
Endnotes, 97, 122
Errors, logical, 47–48
Estimation, 62
costs/benefits, 149
procedures, 147
Evidence, empirical, 47–49
Executive summary, 162
Experience, 43
Explanation, 37

Facts
defined, 51–52
prediction of, 54
Fallacies, logical, 46–47
False cause fallacy, 47
Feynman, Richard, 17
Figures
list, 163
use of, 166, 167
Findings
duplication, 114
implications, 76
research report, 164
writing, 165–167
Fit, 147
Footnotes, 97, 122, 126
Formalist deductivist, 6
Friedman, Milton, 60
Funding
competitive, 232–235
dedicated, 231–232
federal, 231, 232–235
research, 91–92, 231–235
research proposal, example,
177–188

Generalist inductivist, 6
Generalities, new, 75
Generalizations, 53
General Theory of Employment, Interest, and Money, 92, 132
Goals, research, 28–29
Goodness, 63–65
Government agencies, research proposal, example, 177–188
Graduate programs
    creativity in, 31–32
    research methodology, 5–7
Grant proposals, 232–235
Graphs, 97
    use of, 166

Hatch Act funds, 232
Hudson, Darren, 177
Hypotheses, 132–136
    defined, 54, 132
    diagnostic, 54–55, 134–135, 139
    identification of, 54
    maintained, 54, 134, 139
    prescriptive, 73
    problems developing, 135–136
    qualitative, 133–135, 138
    quantitative, 54, 133
    remedial, 55, 135, 139
Hypothesis testing, 54, 132–136

Ideas
    acknowledging, 162
    organizing, 95
    referencing, 121
Imagination, 31
Indexes, 116–117
Induction, 45–46, 56
    scientific approach, 74–77, 78
    statistical, 75
Information
    confidentiality, 152
    organizing, 95
Input/Output model, 92, 149
Institutionalists, 66–67
Integer programming, 149
Intellect, 19

International Monetary Fund, 151
Internet, referencing sources, 124
Interpretation, normative, 74
Interviews, reference format, 123
Introduction
    research proposal, example, 179–181
    research report, 163
Intuition, 30–33, 44
Issues, 112

Jargon, 93, 95
Johnson, Glenn, 9, 12, 59, 102
Johnson, Harry, 60
Journal articles, 160–161
    reference format, 123
Journals, publishing in, 70
Justification section, research proposal, 87

Keynes, John Maynard, 60, 92, 101
Keynes, John Neville, 60
Key words, literature search, 117–118
Knowledge
    accumulated, 37
    descriptive, 56, 66
    disciplinary, 24
    new, 16
    normativistic, 40–42, 55, 63–66, 68–71, 77
    obtaining, 43–45
    positivistic, 40–42, 55, 60–63, 65–66
    pragmatic, 66–67
    prescriptive, 41, 65, 66, 67
    private, 42–43, 55, 63
    public, 42–43, 55, 63
    reliability, 16, 41–43, 45–46, 55
    of values, 41

Ladd, George W., 58, 78
Language, subjective, 107
Laws, causal, 52
Leontief, Wassily, 60, 72, 92
    Input/Output model, 149
Limitations, research proposal, example, 187
Linear programming, 148
Line spacing, 96

Literature
  peer-reviewed, 115–116
  search, 17–18, 115–119, 125
Literature review
  dissertation proposal, example, 212–218
  purpose, 113–115
  quotations in, 120
  research proposal, 87–88
  research report, 163
  sources, 115–116, 125
  thesis proposal, example, 194–196
  writing, 119–120
Little, Brown Handbook, The, 96
Logic
  deductive, 32, 45, 56
  inductive, 45–46, 56
Logit analysis, 147

Macy, Anne Marie, 201, 202
Manuscripts, unpublished, reference
  format, 123
Marshall, Alfred, 60, 132
Master's thesis proposal, example,
  290–299
McCloskey, Donald, 12
Measurement, 44, 71
Methodology
  defined, 4, 25–26
  described, 25–28
  economic, 4, 26–27
  research. *See* Research methodology
Methods
  defined, 4, 25–26, 140
  descriptive, 147–150
  empirical, 142–144, 146–150
  historical, 147
Methods and procedures
  approach to, 153–154
  data, 150–153
  dissertation proposal, example,
    223–226
  empirical methods, 142–144, 146–150
  models, 144–146
  purpose, 141–142
  research proposal, 88
  research proposal, example, 186–188

research report, 164
thesis proposal, example, 197–199
writing, 165–167
Methods of Agreement and Difference,
  142–144
Mill, John Stuart, 142
Models
  econometric, 145
  economic-engineering, 149
  empirical, 145
  Input/Output, 149
  optimization, 145
  purpose, 145
  referencing, 124
  simulation, 145
  undocumented, 126
Monographs, 161
  reference format, 123

National Academy of Sciences, 6
National Bureau of Economic Research,
  161, 231
National Science Foundation, 91
Negative Canon of Agreement, 142–144
Neutrality, scientific, 50
Nonlinear programming, 149
Normativism, 63–66, 68–71, 77
  conditional, 78–79
  objective, 79
Notation, mathematical, 97
Notes, reference, 118–119

Objectives
  dissertation proposal, example, 212
  identifying, 76
  modifying, 90
  research, 28–29
  research proposal, 87, 108–109
  research proposal, example, 181
  specification, 73
  thesis proposal, example, 193–194
Objective statement, developing, 109–112
Objectivity
  in economic research, 62–63
  and funded research, 235
  personal, 49–50

Observations, reliability tests, 47–49
Oldenburg, Henry, 169
Optimization models, 145, 148–149
Ordinary Least Squares (OLS), 147
Orientation, 53
Outcomes
  generalized, 75
  goodness/badness of, 63–65
  reliability tests, 47–49

Page margins, 96
Paraphrasing, 121
Patterns, 147
Phenomenon, and condition, 142–144
Philosophy, and research methodology,
  58–60, 77–78
Plagiarism, 121
Planning, conceptualizing, 128
Platitudes, 95
Popper, Karl, 50
Positive Canon of Agreement, 142–144
Positivism
  logical, 60–63, 64–65, 68–71, 77
  tests of, 63
Post hoc fallacy, 47
Pragmatism, 50, 66–67, 68–71, 78
Precision, provision of, 53–54
Prediction, 37
  conditional, 51
  scientific, 50–55, 57
Premises, 74–75
Presuppositions
  pragmatic, 67
  untested, 61
Principal investigator (PI), credentials,
  86, 188
Principles of Economics, 132
Probabilities, modeling, 150
Probit analysis, 147
Problematic situation, 104–105
Problem(s)
  action, 100, 101
  classification, 101
  decision, 100, 101–102
  defined, 100
  disciplinary, 101

general, 104
  identification, 73, 76, 102–104
  research, 101–102
  researchable, 104, 105, 106
Problem statement, 105–107
  developing, 109–112
  dissertation proposal, example, 203–212
  research proposal, 87
  research proposal, example, 179
  thesis proposal, example, 192–193
Procedures, defined, 140
Propositions
  reliability tests, 47–49
  simple, 132–133
Pseudoscience, 50
Publications
  peer-reviewed, 115–116, 168–170
  popular, 116
Publishing
  journal, 70
  research, 169–170

Quantification, 44, 71
  and positivism, 61–62
Questionnaires, 152
Questions, research, 28–29, 112
Quotations, 120
  referencing, 121

Rationale, research proposal, example,
  181–186
Realities, observed, 75
Reasoning, 44
  deductive, 74–75, 77
  fallacies, 46–47
  inductive, 75–77
References, 120–124, 170
  list of, 122, 164
  notes, 97, 118
  parenthetical, 122
  research proposal, 88–89
Regression, 147
Relationships
  data and, 74
  generalized, 53
Reliability, tests for, 47–49

Reports
  components, 162–165
  technical, 161
  types of, 160–162
Request for Proposals (RFP), 233
Research
  analytical, 24–25
  applications, 18–19
  applied, 20
  basic, 20
  classifications, 19–24
  confirmation, 114
  defined, 15–19
  descriptive, 24–25
  design, 29
  disciplinary, 21–22, 68, 160
  duplication, 114
  economic, 53–54
  funding, 91–92
  interdisciplinary, 57, 234–235
  justification, 102
  library, 116
  misuse of, 12–13
  multidisciplinary, 57, 234–235
  plan, 84, 85
  prior, 114
  problem. *See* Problem(s)
  problem-solving, 23–24, 67, 69
  process, 18, 28–33
  productivity, 70
  reporting, 159-172; *See also* Reports
  subject-matter, 22–23, 68–69
  tools, 148–150
  types, 11
Research methodology, 4–5, 26–27
  empiricism in, 71–72
  flaws, 7
  philosophy and, 58–60, 77–78
  training in, 5–7
Research Methodology for Economists, 9
Research proposals, 84–85
  conceptual framework, 127–138
  elements of, 85–89
  evaluation criteria, 89–90
  example, 177–188
  methods and procedures, 140–154

  objectives, 108–109
  problem statement, 105–107
  review of literature, 113–125
  revisions, 90–91
  schedule, 187–188
Results
  disseminating, 29
  generating, 29
  goodness/badness of, 63–65
  interpretation, 72–73, 74
  preliminary, 169
  reporting, 164
  validity, 159
Revelation, 44
Review
  grant proposals, 232–233
  of writing, 96
Review of literature. *See* Literature review
Revisions, research proposal, 90–91
Rewriting, 93
Risk programming, 149
Rogers, Will, 17

Sampling, 71
Schultz, T.W., 92
Science
  defined, 36–38, 55
  physical, 38–40
  social, 38–40
Scientific approach (method), 72–74
  deduction/induction in, 74–77, 78
Seemingly Unrelated Regression
    (SUR), 147
Senses, 43
Sharp, Shombi, 191
Significance
  research proposal, example, 181–186
  statistical, 145
Simon, Herbert, 20
Simplifications, 53
Simulation
  mathematical, 149
  models, 145, 149–150
  probalistic, 150
Simultaneous equation models, 147
Situation, problematic, 104–105

Software, reference format, 124
Sources
  citations, 118
  data, 150–151
  dissertation proposal, example, 226
  documentation, 88–89
Special pleading fallacy, 46
Statistics, tools, 147–148
Stochastic programming, 149
Subjectivity, 49
Summary, 53
  reporting, 164
  research proposal, example, 179
Surveys, 152
Symbols, mathematical, 97
Synthesize, ability to, 9–10
Systems analysis, 149

Table of contents, research report, 162–163
Tables, 97
  list, 163
  use of, 166, 167
Tautologies, 79
Testing, 40
Theory
  confirming, 39
  defined, 52–54
  formulating, 130
  function of, 53–54
  refinements, 130–132
  relevance, 131
  test of, 57
Thesis
  authorship, 173
  completeness of, 161
  sample proposal, 190–199
Third person form, 97
Thoughts
  crediting, 121
  organizing, 95
Three-Stage Least Squares, 147

Time-series analysis, 147, 148
Title
  research proposal, 86
  research report, 162
Title page
  dissertation proposal, example, 201
  research proposal, example, 178
  research report, 162
  thesis proposal, example, 191
Trend analysis, 148
Truth, 16–17
Two-Stage Least Squares, 147

Understanding, 61
U.S. Chamber of Commerce, 150, 151
U.S. Department of Agriculture, 232
  Cooperative Research Information
    System (CRIS), 126
Universities
  funding by, 231–232
  land-grant, 232
Utilitarianism, 63–64

Validation, empirical, 40
Values
  and facts, 52
  positivism and, 62
Variables, control of, 144
Vector autoregression, 148
Verb tense, 97

Williams, Willard, 128
Words, judgmental, 95
Workability, 66
  tests of, 48, 67
Writing
  ambiguity in, 95
  critique of, 229–230
  guidelines, 94–97
  importance, 92–94
  style, 96

Printed and bound by CPI Group (UK) Ltd, Croydon, CR0 4YY

23/04/2025

14660906-0002